MEXICO'S VALLEYS OF CUICATLÁN AND TEHUACÁN

The Southwest Center Series

Jeffrey M. Banister, Editor

DAVID YETMAN AND ALBERTO BÚRQUEZ

MEXICO'S VALLEYS OF Cuicatlán and Tehuacán

From Deserts to Clouds

THE UNIVERSITY OF
ARIZONA PRESS
TUCSON

The University of Arizona Press
www.uapress.arizona.edu

We respectfully acknowledge the University of Arizona is on the land and territories of Indigenous peoples. Today, Arizona is home to twenty-two federally recognized tribes, with Tucson being home to the O'odham and the Yaqui. Committed to diversity and inclusion, the University strives to build sustainable relationships with sovereign Native Nations and Indigenous communities through education offerings, partnerships, and community service.

ISBN-13: 978-0-8165-4873-6 (paperback)
ISBN-13: 978-0-8165-4874-3 (ebook)

Cover design by Leigh McDonald
Cover photo: The town of Quiotepec, Oaxaca, near the confluence of the Río Grande and the Río Salado. Photo by David Yetman.
Designed and typeset by Sara Thaxton in 10.5/15 Warnock Pro with Bulmer MT Std

Unless otherwise noted, all photographs and diagrams are by the authors.

Publication of this book is made possible in part by financial support from the Instituto de Ecología, Universidad Nacional Autónoma de México and the Southwest Center of the University of Arizona.

Library of Congress Cataloging-in-Publication Data
Names: Yetman, David, 1941– author. | Búrquez, Alberto, author.
Title: Mexico's valleys of Cuicatlán and Tehuacán : from deserts to clouds / David Yetman and Alberto Búrquez.
Description: Tucson : University of Arizona Press, 2023. | Series: The Southwest Center series | Includes bibliographical references and index.
Identifiers: LCCN 2022061320 (print) | LCCN 202206 (ebook) | ISBN 9780816548736 (paperback) | ISBN 9780816548743 (ebook)
Subjects: LCSH: Natural history—Mexico—Reserva de la Biósfera Tehuacán-Cuicatlán. | Human geography—Mexico—Reserva de la Biósfera Tehuacán-Cuicatlán. | Reserva de la Biósfera Tehuacán-Cuicatlán (Mexico)—Geography.
Classification: LCC F1210.9 .Y48 2023 (print) | LCC F1210.9 (ebook) | DDC 972—dc23/eng/20230113
LC record available at https://lccn.loc.gov/2022061320
LC ebook record available at https://lccn.loc.gov/2022061321

Printed in the United States of America
♾ This paper meets the requirements of ANSI/NISO Z39.48-1992 (Permanence of Paper).

CONTENTS

ACKNOWLEDGMENTS

Numerous individuals assisted us during the preparation of this book. We especially acknowledge assistance of Pedro Torres for transportation and guidance on Oaxacan culture, many colleagues at the Universidad Nacional Autónoma de México—Instituto de Biología, Instituto de Ecología, Instituto de Geología, Facultad de Ciencias, and Facultad de Estudios Superiores Iztacala. Patricia Dávila resolved questions related to the flora of the valleys; Abisaí J. García-Mendoza, Luis Eguiarte, and Greg Starr helped with identifying *Agave* species; and Mariano Elías-Herrera and Thierry Calmus revised the chapter on geology, which has proved most difficult. Luis Arroyo, of Casa de los Abuelos in Oaxaca, shared his vast knowledge of the workings of Oaxaca, and Salvador Chávez, of San Gabriel Casablanca, introduced us to Cerro Petlanco. Thanks to many anonymous informants in small towns and remote locations, who directed us to a host of unusual places and helped us to find rare plants, shelter, and food in several dire situations (in that order). The Southwest Center, University of Arizona, and the Instituto de Ecología, Universidad Nacional Autónoma de México, provided freedom and support for the authors. We thank Lynn Fowler and Angelina Martínez-Yrízar for their continued support and suggestions. And we offer thanks to anonymous reviewers, who forced us to sharpen our focus and eliminate potentially egregious errors.

MEXICO'S VALLEYS OF
CUICATLÁN AND TEHUACÁN

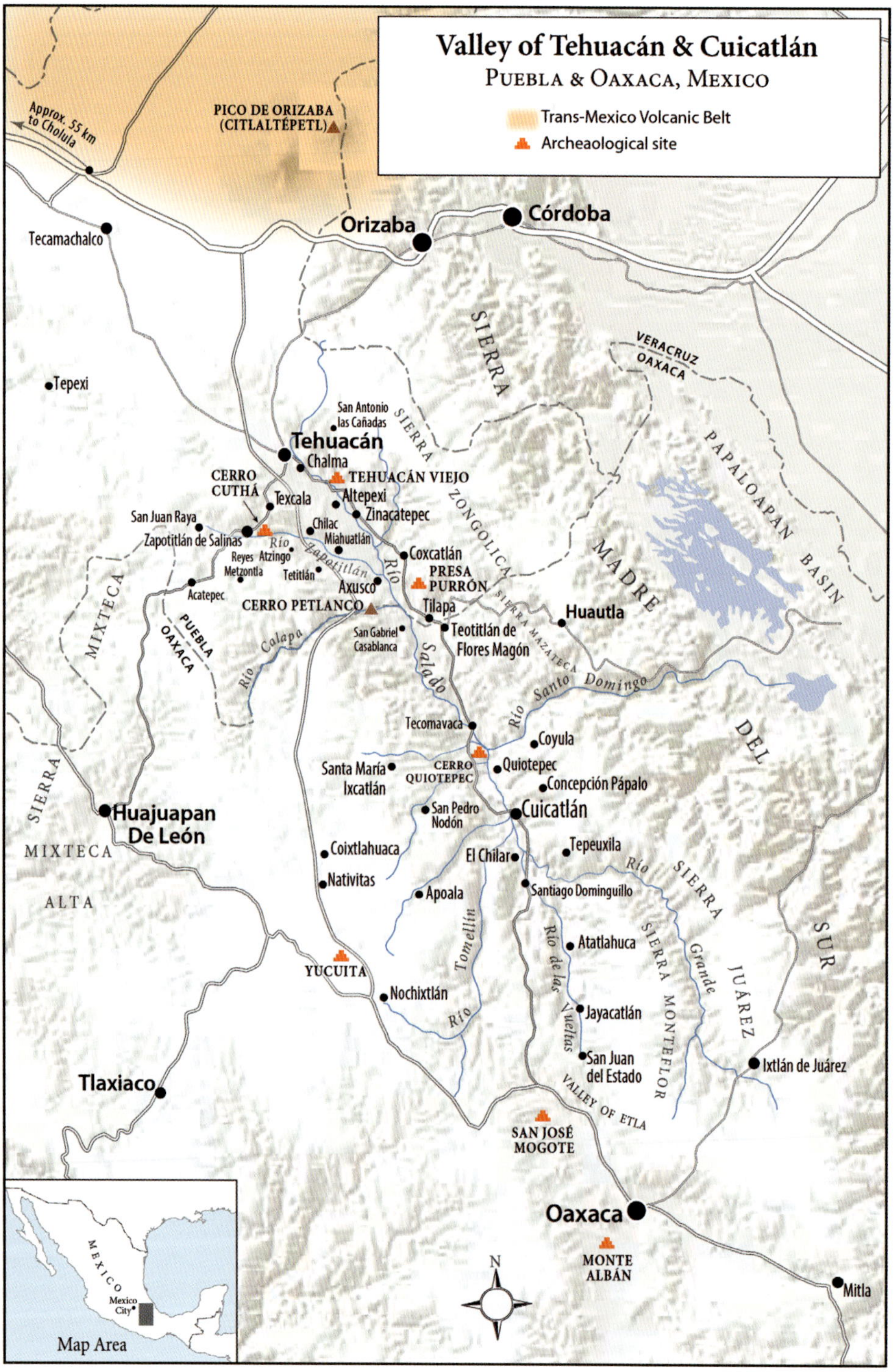

MAP 1 Map of the Valleys of Tehuacán and Cuicatlán and their location in Mexico. Map by Paul Mirocha.

Introduction

Why We Chose These Valleys

I first traveled through the Valley of Tehuacán in 1969, driving an old Land Rover south en route to the city of Oaxaca. I recall very little of the region, except for a semidesert landscape, unending mountains, interminable curves, and the plodding, smoking diesel trucks crowding the narrow, shoulderless highway. Those trucks, known in Mexico as *tórtones*, heavy, usually overladen, are seldom seen now. In those days, *tórtones* clogged the mountain roads, belching black clouds of diesel smoke. Their parking brakes would often fail, so when drivers suffered a flat tire, they would block the wheels with large rocks to keep the monsters from rolling out of control. The tire replaced, the operators would drive away, leaving the large rocks behind for other vehicles to run into. My Land Rover was a good choice for that terrain, for the road was also laden with potholes, cracks, washouts, and landslides.

A modern expressway connecting the cities of Tehuacán and Oaxaca would not be completed until after the turn of the twenty-first century. The road through the Cuicatlán Valley, which connects to the Valley of Tehuacán and leads nearly to Oaxaca, was still a dirt track. It often washed out during the summer rains or was rendered impassable by multiple landslides. If paved, that route would have shortened the trip by a couple of hours.

It was not until the year 2000 that I visited the valleys themselves, walking through the hills and stopping by some of their small towns. By then the roads had been expanded and improved, and graded roads replaced many unimproved tracks. Since that trip I have logged more than twenty visits, discovering sights, peoples, and natural history features I had previously overlooked. Potholes are now fewer. Road-blocking landslides are still a hazard.

Alberto Búrquez joined me on an exploration of the valleys in 2003. He had visited previously as a lecturer in ecology at UNAM, Mexico's National Autonomous

University. He could hardly resist bringing his students to one of the world's biodiversity hotspots. I have often envied the easy access he had in those years to the valleys of Tehuacán and Cuicatlán, only a few hours' drive from the southern limits of Mexico City, where UNAM is situated.

Alberto and I collaborated on projects throughout the 1990s, focusing on the plants and vegetation of the northwestern Mexican state of Sonora, where he had been born and raised and now held a distinguished post as professor of ecology for UNAM. We had a list of places in Sonora to visit and study, he as an ecologist studying the relationships between plants and animals, I as a philosopher who found Mexico's natural and cultural history so rich that I abandoned my philosophical musings. On another visit to Oaxaca around 2005, we agreed that it was time for us to collaborate on a book on the valleys we had come to hold in the highest esteem. Though they lie far from the Sonora with which we are familiar, the valleys bear a close ecological resemblance to that state far to the northwest. Out of that easy agreement with my friend came this book.

What would impel us to expend the effort and expense in writing a book about these valleys, so far from our homes in the Sonoran Desert? After all, the first impression visitors experience for much of the year is one of semidesert, drought, and, in places, a parched, often eroded landscape (except after summer rains). Yet unless one is in a hurry to get from Tehuacán to Oaxaca or the reverse, it is difficult not to be impressed by the vegetation and landscapes visible from a vehicle. The combination of cactus forests, plants of unusual shapes and densities, and minor roads leading off into the bush and hills in all directions poses an irresistible draw to anyone with a curiosity about natural and human history. The mountains on either side that engulf the valleys seem to shield mysteries beyond the cliffs and forests that ring the east side and the forbiddingly steep desert slopes on the west. The landscapes away from the cities and viewed up close reveal human occupation deep in antiquity. Churches, ancient as well as new, most of them visible from afar, grace every settlement, be it a village or a town. Place names roll off the tongue, evoking times long before Europeans ordered the prefacing of aboriginal names with the titles of saints, Indigenous names like Alpizagua, Atatlahuca, Altepexi, Atolotitlán, Axuxco, Coxcatlán, Metzontla, Miahuatlán, Nanahuatípam, Tecomavaca, Teotitlán, Zapotitlán, Zinacatepec, and on and on. The modern, urban Mexico of the city of Tehuacán grades quickly into hamlets and villages, where old traditions endure and life proceeds at a slower pace. Sophisticated dwellers from the megapolis of Mexico City find the allure of the valleys as compelling as I do.

The closer we looked, the more extraordinary and complex the valleys became. Dense forests of columnar cacti swathe the hillsides with their color: there are eighteen species of the giants, more than in any similar tract in the world. Within

FIGURE I.1 *Pachycereus weberi* and thornscrub, Santiago Dominguillo, Valley of Cuicatlán.

the valleys we find not just unusual vegetation, but also a host of endemic species and strange plants with names like elephant's foot, mother-in-law's chair, old man, and (ahem!) ball swellers. The endless varieties and combinations of trees, shrubs, agaves, yuccas, and cacti poke out of cliffs, protrude from tropical forests adjacent to barren deserts, emerge from unexcavated pyramids, lurk in obscure canyons, and hide in oak woodlands and pine forests. The Indigenous peoples of the valleys, at least eight different linguistic groups, persist, some even thriving.

All this is a result of the landforms, the geological underpinnings that account for the region's multiple climates and habitats as well as its evolutionary isolation. Valleys ultimately end in flatlands, of course, but these valleys meet and form a gorge before flattening out, introducing the flora of two different tropical realms to each other and bridging a gap between two landmasses. Here is the connection, the artery, the corridor, the bridge tying the landmass of Central America and Mexico's great Central Plateau together, a route where peoples have routinely crossed from one landmass to the other for two hundred generations, and a route where living organisms from the north and south mix and send out emissaries in all

FIGURE 1.2 Church and plaza, San Luis Atolotitlán, Valley of Tehuacán.

directions. The valley bottoms themselves are remnants of an enormous ancient lake that bequeathed deep soils to future agriculturalists. The steep slopes to the east and west represent the workings of vast faults, mostly triggered hundreds of millions of years ago. Stupendous mesas of pure limestone, their innards honeycombed with passages, hover over canyons. The mountains capture rain and guide it to the fields below. Everywhere, irrigation works, some barely discernible, bespeak the labor of centuries, as farmers struggled to usher that water to cornfields.

And on all sides are the monuments of human origin, the fragments and monoliths of thousands of years of human history, well concealed, but once revealed, declaring that these valleys were worth ruling, worth fighting for, worth irrigating, worth elevating to international prominence. Slowly, painstakingly, archaeologists have coaxed the chronicles of humanity from the valleys, revealing a past of obscure human origins ten thousand years ago to the glorious temples and residences of Mesoamericans of Classic and Postclassic times. The great structures have begun to emerge from their camouflage of millennia of dust, of shrubs, vines, and trees, disclosing the high value that pre-Columbians placed on the valleys as a place to call home.

As Alberto and I became aware of layer after layer of complexity and distinctness, we agreed we needed to present to the world the glory, the variety, the mosaic of landforms, climates, habitats, and plants, and the many peoples of varying ethnicities and their achievements that make the Valleys of Cuicatlán and Tehuacán as fascinating as any such areas on Earth. That is what we hope to achieve in this work. We welcome you to Mexico's great connecting bridge, the Valleys of Cuicatlán and Tehuacán.

Herein we describe what we have seen and learned about these valleys, what we hope is a comprehensive geographic chart, with special emphasis on plants. We do not offer new research, only insights from our travels and from studies done by others with a sharper focus. Our description of what we have learned begins in chapter 1 with a general description of the geographic orientation of the valleys. We follow this in chapter 2 with an introduction to the valleys' climate and hydrology, that is, the confusing variety of climates in the valleys and why they constitute a zone of aridity within the tropics. The climate, in turn, reflects the influence of the valleys' often bewildering geology, to which we turn in chapter 3. We describe how the topography of the region was shaped over deep historical times and why the valleys appear as they do in the present. In chapter 4 we present a lengthy sketch of the prehistory of the valleys and why they have assumed an important role in the deep and contemporary human history of Mesoamerica. In chapter 5 we describe the cultural and linguistic background of the valleys, how various peoples have interacted over the millennia to produce the cross section of humanity we find there today. In chapter 6 we focus on four archaeological sites that encapsulate the valleys' complex social and material history prior to the European Conquest in 1521, locations inadequately recognized in public awareness and popular archaeological literature of Mesoamerica. To visit three of these sites, one must make local inquiries. One of them, Tehuacán El Viejo, also called La Mesa, could not be visited at all until recently. It is now a celebrated and monumental site.

The plants are what have attracted the most international attention to the valleys. Our often-detailed discussion of the plants and the associations they form constitute chapter 7. Finally, we conclude in chapter 8 with a brief presentation of Cerro Petlanco, an inconspicuous hill near the intersection of the two valleys. There we find combined biology, geology, culture, and history that epitomize the valleys' role as a Mexican Biosphere Reserve.

We wish to insert one note of caution. We have attempted to present the most current findings in our discussions of the history of ancient sites and peoples in the valleys. Archaeological investigations have expanded dramatically in recent decades, increasing our understanding of the prehistory of the region at a pace that outstrips our ability to describe the most up-to-date accounts. While we present what we believe to be the best information available, it is subject to revision at any time, and we note that experts often disagree on the accuracy of historical descriptions, sometimes vociferously.

—*David Yetman*

I remember my childhood in the desert hills and arroyos of the Sonoran Desert. I harvested sweet and red pitaya fruits and searched the desert for prized flint arrowheads and eggshell pottery of ancient Seri Indians. At home, the family library had hundreds of volumes, most of them in Spanish and French, but also a sizable collection of works in English and German. I spent hours going through the books, some clearly forbidden for my age. At the time, adventure books were my choice. They provided valuable information about the wilderness, like the unique species of the arctic tundra and the plants of the Malaysian archipelago. Others offered more practical advice, like survival skills in the desert and improvisation of fishing gear, and fun, which included building boats of Styrofoam, making black powder, or exploring caves with smoky torches that, once inside the dark underworld, extinguished and almost choked us to death. A happy childhood, wandering in the desert and at the seashore, discovering the pleasure and value of reading, marked my life.

That early provincial training did not prepare me for the startling environmental revelations of college life at the National University of Mexico in Mexico City. Here, at the top of the Mesoamerican world, in Carlos Fuentes's region, where the air is clear, I found an astonishing metropolis surrounded by unending wonders of nature, two thousand kilometers south of my birthplace, two thousand meters higher in elevation, and as alien to a Sonoran citizen as Tibet to an English person. The mighty mountains and forests of the Trans-Mexican Volcanic Belt (TMVB) were a new and mysterious realm of clouds, summer coolness, water, and greenness, a stark contrast to the open desert I was leaving behind. Once there, I explored glaciers in the high volcanoes, the dark, forbidding jungles of the humid lowlands of the Gulf of Mexico and the Guatemalan Petén, and the slightly more familiar dry tropical forests of the Pacific coast.

As is the case in other large modern countries, Mexico is not a homogeneous entity; it is more a jigsaw puzzle of interlocking ecological and cultural units that are, for the moment at least, united under a single central government. The dry northwestern states, characterized by arid plains along the coast, a few fast-flowing short rivers, and deep canyons in the Sierra Madre, have little in common with the flat, calcareous Yucatán Peninsula, in which all rivers are subterranean and the few isolated peaks, such as the Cerro de Xal, jutting up in the Puuc region, are lower than 250 meters. The Mexican central highlands have more in common, climatically speaking, with forests of northern latitudes than with the nearby luxuriant tropical lowlands of the Gulf of Mexico and the Pacific. The ethnic and cultural composition changes dramatically from region to region, in some cases from valley to valley.

Pockets throughout Mexico reveal a strong European influence, while others, like the Oaxacan highlands, retain the ancestral pre-Columbian genetic and cultural makeup veiled under a thin Western veneer.

Coming from Sonora allowed me to view Mesoamerica as an outsider. I understood that Mexico City, the capital of the mighty Aztec soldiers, was rich and so strategically located as to exert political and economic control over its surrounding polities. From there, the Mexica conquered their neighbors and sent punitive expeditions to keep goods flowing. I also had a firsthand understanding of the plight of a Mexican society torn between the revered ancient mythology and the search for contemporary modernity, a conflict almost absent in the northern reaches of Mexico where I lived. According to José Vasconcelos, a famed but somewhat cynical scholar, politician, and philosopher, the Northwest was the land where culture ends, and grilled beef begins. As I absorbed the history and culture of the Mexico basin, I gradually left behind the romantic idea of benign, supposedly peaceful Mesoamerican cultures and started to understand the more realistic interplay between statelets and kingdoms that, at the time of the Europeans' arrival, were in an almost continuous state of warfare, spoke a host of mutually unintelligible languages, and emerged from contrasting cultural backgrounds.

What I did not realize at the time was the role of geography in isolating not only animals and plants but cultures as well. The mighty mountains of central Mexico at the TMVB and the Sierra Madre del Sur represented then and now a formidable barrier to human movement and communication. At the time of the Europeans' arrival to Mexico, five hundred years ago, there was a dense network of roads and commercial routes. Contrary to the popular perception that Native Americans lived as hunter-gatherers immersed in a nomadic way of life, millions of natives lived in prosperous urban centers with straight paved avenues, sophisticated hydraulic systems for agriculture, schools, and public sports competitions like *tlachtli*—ball games.

Roads among these thriving cities allowed the long-distance movement of goods and people. These thoroughfares were also migratory, political, and ideological corridors. Spanish conqueror Hernán Cortés confined his route to secondary paths, enabling him to forge alliances with local chiefs and vassal statelets, seek cover from potential ambushes, and keep as low a profile as possible. Most of his journey to the capital city of the Mexica was parallel to the well-trodden "highway" that linked the coast with the highlands. Cortés and his followers were astonished to learn that fast relay runners carried exotic goods like ice and snow or fresh fish between the Gulf of Mexico and the capital. He wrote to the Spanish king that in about one day, couriers informed the Aztec nobility of the Spanish presence, even though the abrupt geography and the lack of carriages and pack animals imposed severe limitations to travel. Until recently, with the advent of freeways, modern tunnels, and high bridges, moving between relatively close places in central Mexico took hours, even days. At the time of Cortés's army, the manageable way to travel was confined to a few demanding natural passages.

The most convenient route from Mexico to Oaxaca (and southeastward to Chiapas and Guatemala) was through the gentle descent of the natural ramp of the Valley of Tehuacán and the steep but short uphill staircase along Río de las Vueltas and some of its tributaries at the southern end of Cuicatlán Valley, a trip flanked by the Mixtecan mountains to the west, and the Sierra Madre del Sur to the east. This route was open enough to prevent ambushes and plentiful in crops and orchards capable of sustaining the weary traveler or the marching army. The idea of the ramp, along with the singularly transfixing landscapes, strange flora, and outstanding cultural milieu, was a natural avenue to start this book that my friend David Yetman proposed and graciously invited me to co-author.

During my childhood in the 1960s, lemonades and highballs were always prepared with the famous mineral waters of Tehuacán. The soda water carried a label on which the National University of Mexico's Institute of Geology certified and detailed the chemical content of the reputedly miraculous waters. Little did I know that someday I would visit the Valley of Tehuacán and learn about the vegetation, peoples, and dwindling springs of the valley. Many years later, my journey took me to the heart of the valleys, to places like Purrón, Coxcatlán, and Petlanco, where the earliest Americans roamed, and where kings from faraway places came looking for the sacred water, in search of a cure in secluded travertine springs. In the numerous treks David and I took throughout the valleys, we seemed to sense the spirits of earlier explorers: Karl Albert Purpus, the intrepid Dutch German plant collector; Cassiano Conzatti, the indefatigable Italian émigré who opened Cuicatlán to the plant sciences; novelist Malcolm Lowry drinking mezcal in the lost railroad town of El Parián; Richard S. MacNeish and many extraordinary researchers who have documented the natural history and human occupation of the valleys. I can imagine all of them sipping the bubbling water from the Axoxopan springs before starting their journey.

My first visit to the valleys was as an undergraduate student back in 1974. I remember entering the city of Tehuacán, imbued with the romantic idea of the mineral water springs, with almost no knowledge about the valleys' renowned flora, and even less about the extraordinary cultures that populated the valleys and their

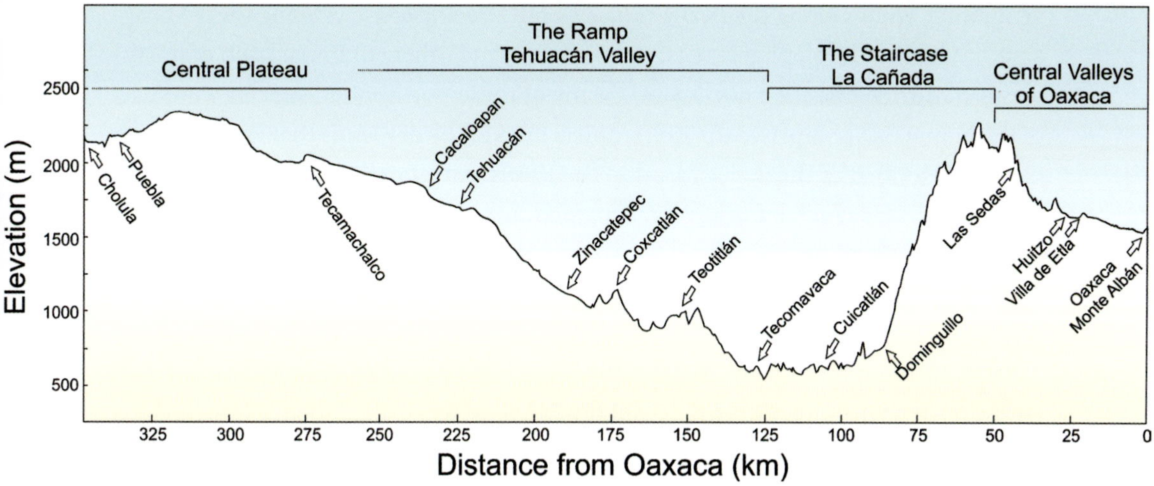

FIGURE I.3 Elevation profile along Hwy 150 from Puebla to Tehuacán and Hwy 135 from Tehuacán to Oaxaca, illustrating the gradual descent into Cuicatlán and the rapid climb toward Oaxaca City.

surroundings. The instructors took us through a winding road toward Zapotitlán Salinas. The road slowly revealed a landscape similar to my Sonoran Desert homeland, but wilder, richer, more opulent in charismatic plants. The columnar cacti covered the mountainsides by the millions, and their diversity in size and form was astonishing. Agaves of all sorts, along with massive globose cacti and enormous *Beaucarnea* "palms," which resembled more a brontosaurus's than an elephant's foot (as locals called them). All these fantastical species emerged from a whitish calcareous terrain, where sometimes ancient fossil seashells paved the paths.

Later, I made many visits to the valleys, usually just passing by on my way toward the lowland tropical forests of Veracruz, Campeche, Tabasco, and Chiapas, or when taking students for ecology class fieldwork, to give them a glimpse of the landlocked arid lands of the valley. I was always impressed by the environmental change from the stark, arid northern reaches of the Tehuacán Valley—where the road, built on the old pre-Hispanic road from Mexico City to Veracruz, descended dramatically eastward in the short distance of about six miles—to the verdant, luxuriant forests of the high sierra at Cumbres de Acultzingo, from where I had unbroken views across the desert valley, so reminiscent of my homeland.

For many people the Tehuacán and Cuicatlán Valleys are harsh, intimidating territories. For a Sonoran Desert native, however, these valleys are familiar ground. Indeed, here in these isolated patches of desert, the soul of the Sonoran Desert evolved over the eons. This occurred during the drying trend of the Tertiary period, when plants and animals from these valleys dispersed, following intricate geographic routes along the *tierra caliente* of the Pacific coast. Among these voyagers, the jaguars, the elephant trees, the saguaros, and other columnar cacti of northwestern Mexico marched northward. Many of these northbound species were part of the biome that comprises the rich tropical flora of the dry lands that Howard Scott Gentry described as short tree forest in the 1940s. As rainfall became scarcer to the north, these dry lands naturally transformed into the Sonoran Desert.[1] Now, with ever-increasing global temperatures, some old denizens of Tehuacán have already taken hold in the southwestern United States and promise to venture even farther north in fits and starts.

High in the mountains of the southwestern flank of Tehuacán Valley, in stark contrast to the dry valley bottom, lived the Mixtecs, who called themselves the Ñuu Dzahui, people of the rain, later called, by the invading Nahua, Mixtecah (cloud people). They have remained in those highlands, while other native cultural groups populated the valley bottoms as well as the high mountains. Many from these ancient groups are also now moving from their ancestral lands to the north—migrating in search of more opportunities, bringing with them their culture and their genes.

In the 1990s, I also moved north, leaving the Mesoamerican capital to start research programs in the Sonoran Desert, the northern tropical dry forest, and the high Sierra Madre Occidental. In 1992 I met David Yetman. He invited me to join our friends Paul S. Martin and Cynthia Lindquist on the faculty of Tucson Audubon Society's Institute of Tropical Ecology, held in Alamos, Sonora. Immediately, I discovered a complementary academic colleague. Since then, we have embarked on many enterprises, from discovering the haphazard transport and dispersal of columnar cacti seeds by the hunter-gatherer Comcáac to studying the effects of changes in the Mexican Constitution leading to privatization of communal land. All this while David was filming his TV program, *The Desert Speaks*, on which I was often invited. In the hard-academic arena, however, David and I were following our own widely separated careers; in my case, papers in population ecology and genetics, while David was more into the social and botanical aspects of natural resource distribution, uses, ethnobiology, and history. This book arose as a natural outcome of our formal and informal ongoing collaborations. It is based on conversations and ideas of projects exploring the link between social sciences and biology, always seeking the connection among ecological dynamics, biodiversity, and native peoples and their use of natural resources. Now, with a bottle of Tehuacán water, some mezcal, and sal de gusano, David and I discuss this book in an enchanted town somewhere up in the valleys. Shall the title mention the contrast between the aridity of the valley bottom and the lush forests with dangling Spanish moss of the highlands? Shall it

1. Martin et al. 2000.

mention the Ñuu Dzahui and their sacred relationship to rain and clouds? Shall we go for something like *The Road to Petlanco*?

We ended with a more sober descriptive title and then drove our conversations with the need to close gaps in our information and to incorporate the parade of images that appear in this final version. The title conveys perfectly what we did but inevitably fails to capture the magic of the discovery and the surprise of meeting people from cultures new to us in places we had only known as points on a map. We sip the mezcal, look at the sunset looming over the valleys, enjoy each other's company and that of newfound friends, and speculate about additional places we would love to visit and write about, as always, dreaming of new challenges and adventures.

—Alberto Búrquez

CHAPTER 1

The Mexican Landscape

A Brief Synopsis of the History and Prehistory of the Valleys

It is probable . . . that this region had been a cultural passageway in prehistoric times, both between lowland Veracruz and highland Oaxaca-Puebla, especially the Oaxaca-Cañada-Tehuacán-Cholula-Valley of Mexico chain.

—EVA HUNT, "IRRIGATION AND THE SOCIO-POLITICAL ORGANIZATION OF CUICATEC CACICAZGOS"

When Hernán Cortés and his band of Spaniards set foot on the Mexican landmass near Veracruz in 1519, they marveled at the convoluted topography that confronted them. From their ships anchored in the Gulf of Mexico, they caught glimpses of the towering volcano Citlaltépetl (Mount Orizaba, Pico de Orizaba), more than 5 kilometers high. Soon they would learn to their dismay that to challenge the rule of the Aztecs of Tenochtitlán, now Mexico City, they would be forced to move their armies, horses, and cannon over mountain ranges with ridges that exceed 2,400 meters (7,874 ft.) and passes that reach as high as 3,600 meters (11,811 ft.) in elevation.[1]

Mexico City now dominates the Basin of Mexico, which has been home to the heart of Mexican civilizations for at least the last couple of millennia. It sits at 2,240 meters (7,349 ft.) above sea level and is surrounded by towering mountain ranges on three sides. Only from far to the north can it be approached over land that is not torn and elevated to great heights by uplift from tectonic forces and volcanic eruptions, thrusting and tearing the Earth's surface.

Images from space reveal the bulk of the Mexican landmass—and Mexico's most heavily populated area—compressed among three vast ranges: the Sierra

1. That same topographical morass has proved an immense challenge over the last two centuries to Mexican leaders struggling to build a nation. The centrifugal tendencies among Mexican peoples have been magnified rather naturally by the looming presence of the nation's massive mountain ranges and the intervening canyons, ridges, cliffs, and lava flows.

Madre Oriental (the first major range that Cortés and his army climbed) on the east, the Sierra Madre Occidental on the west, and the Sierra Madre del Sur on the south. In between these three major ranges is the Trans-Mexican Volcanic Belt (TMVB), a youthful and imposing mountain range about 900 kilometers long and up to 300 kilometers wide at its widest point. The bulk of Mexican population lives within the TMVB, an elongated, topographically complex trapezoidal corridor running in an east–west direction. The Sierra Madre Oriental and the Sierra Madre Occidental flank the TMVB and stretch away at parallel angles to the northwest. The northern limit of the TMVB is formed roughly by the Mexican Central Plateau, an area far from the border with the United States, a vast expanse where desert plains and isolated mountain ranges dominate the landscape, and water sources are few.

This string of volcanoes bisects the country at about 19° north and is a manifestation of the subduction by the Cocos and Rivera Plates in the Pacific Ocean under the Mexican landmass. Given its relatively recent geological age, this extensive belt hosts the largest and most of the youngest mountains of Mexico, including a string of about twenty stratovolcanoes, most of them dormant, but two showing eruptive activity. The southern edge of the TMVB, which was formed by eons of explosive volcanic activity and tectonic uplift, presents a formidable barrier to human passage. Cortés encountered its complex topography when military considerations forced him to follow a convoluted path toward the Valley of Mexico, crossing the Paso de Cortés—the pass between the stratovolcanoes Popocatépetl and Iztaccíhuatl at 3,400 meters (11,155 ft.), likely higher than any of the Spaniards had ever stood before.

The Basin of Mexico marks the southernmost part of the vast Central Plateau. With the formation of the TMVB rising in the south, it became a closed drainage basin, or *endorheic basin*, retaining water from streams in the surrounding mountains with no outflow, and hence it was dotted with lakes and permanent or seasonal swamps. The basin extends north between the two Sierra Madres, into the larger Central Plateau, which in turn connects to the Northern Plateau that reaches the United States, and south to its abrupt edge at the TMVB in the immense urban area of Mexico City. Over the centuries that megapolis has drained, pumped dry, or paved over most of the lakes and swamps that once covered the region.

The southern boundary of the TMVB trapezoid is formed by the Sierra Madre del Sur, a more amorphous range, which roughly parallels the TMVB and occupies much of southeastern Mexico. Physiographically, Mexico south of the TMVB can be considered a separate geographic province from the north. The differences are not merely topographical: the climate of the Central Basin to the north is mostly mild to cool and often semiarid, wholly different from much of the climate to the south, which is largely warm to torrid and often wet. Mexico's largest rivers and most of its remaining tropical forests lie to the south and east. There, also, dwells the bulk of Mexico's Indigenous people.[2]

Within the TMVB live the majority of Mexicans, as has been the case for many centuries. That intense concentration of inhabitants is part of a worldwide phenomenon correlating dense population with areas of high volcanic activity.[3] The physical barrier imposed by the TMVB has proved a cultural and historical barrier as well, so much so that to some extent, Mexico can still be viewed population-wise as it was before the arrival of Spaniards. In the heart of the TMVB lies the land of the Nahuas, and to the west lies Michoacán, land of the Purépecha, or Tarascans. The densely populated Basin of Mexico features a cool, semidry climate and soils ideally adapted to growing corn, beans, and squash. North of the basin lies what the Aztecs called the Gran Chichimeca, at that time sparsely populated by largely nomadic peoples whom the Aztecs viewed as barbarians (though acknowledging them as their ancestors). Much, but by no means all, of that portion of Mexico is still sparsely populated. South of the volcanic belt, in the Mesoamerican landmass, Maya, Olmec, and Zapotec civilizations, all with origins nearly four millennia in the past, originated, prospered, and declined, independent for the most part of developments in Mexico during the same period. Teotihuacan, the enigmatic and brilliant civilization of two millennia ago; Tula of the Toltecs, which flowered toward the end of the first millennium of the Common Era; and much later, Tenochtitlán of the Aztecs, all originated in the Basin of Mexico.[4]

Despite the physical separation of the two loci of Mesoamerican civilizations, exchange among cultures was ongoing and influential on both sides of the TMVB. Compelling cultural and geographic reasons have led archaeologists to consider all these cultures, from Teotihuacan to Guatemala, as part of Mesoamerica. For two millennia, the cultures south of the TMVB experienced relative autonomy from Mexico thanks to terrain and physiography. Still, humans have crisscrossed between

2. The Trans-Mexican Volcanic Belt broadly marks the separation between the two biogeographic ecozones of the Americas: Neotropical and Nearctic. Mexico has fifteen physiographic regions. Five of them lie south of the TMVB, and the remaining ten lie north of it. See Rzedowski 1978.

3. About 500 million people live near volcanoes (Doocy et al. 2013, 5). Because of their soil fertility and complex topography, volcanic soils cover 1 percent of the surface yet support 10 percent of the population, including some of the highest human densities (Neall 2009).

4. The correct pronunciation of Teotihuacan, the great city of Classic times, is by no means a settled issue. Purists prefer what they consider a pronunciation close to the Náhuatl, which, they believe, would render the spelling Teotihuacan (stress on the penultimate syllable). Spaniards appear to have preferred the more common Teotihuacán, with the stress on the last syllable. We take no stand on the question but adopt the simpler spelling, Teotihuacan. The same applies to many other place names, except Tenochtitlán, to whose last syllable we choose to apply the accent.

the two landmasses for eons, forming trails and roadways. We were merely latecomers to treks begun thousands of years earlier.

Beginning early in human prehistory, hunters and gatherers established thoroughfares, followed later by the early agriculturalists, later yet by merchants, and late in prehistory by armies of soldiers.[5] Generations of traders plied the great distances among civilizations on foot to exchange their wares, leaving behind them remnants of their imported technology and merchandise as the empires developed, matured, decayed, and vanished. Commercial activity spanned the landmasses long before the Common Era, covering more than 3,000 kilometers from deep in Central America to the Mississippi Valley, perhaps even beyond.[6] Toward the end of the Postclassic period (the six centuries or so from the waning of the great civilizations to the Conquest, roughly 900–1521 CE), great armies, thousands strong toward the end of the period, plodded back and forth, often dragging captives with them, north and south.

The principal connection between these landmasses separated by the TMVB has been a corridor beginning in southern Puebla state and continuing east-southeast into Oaxaca: the Valley of Tehuacán in Puebla and the Valley of Cuicatlán in Oaxaca, also called La Cañada. For travelers in late pre-Columbian Mesoamerica between the Great Chichimec and southern and eastern Mesoamerica, the path of least resistance was through Tenochtitlán or the vicinity of Lake Texcoco, and then bordering the great mountains to the east, passing near Cholula, to either the north or the south to avoid the fierce Tlaxcalteca—a Mesoamerican premodern republic like classical Athens.[7] From Cholula, the easiest route then followed what we call the *ramp* into the Valley of Tehuacán, south into Cuicatlán. From that low point, it climbed up what we consider the *staircase* to the upland valleys of Oaxaca. That route (or routes) avoided rugged terrain and large rivers while mostly traveling in a rather pleasant climate. From Oaxaca, travelers—and there were many—could readily continue south to the lowlands of the Maya in Chiapas and Central America.

The alternative routes were three. One involved a southerly crossing through deep valley after deep valley in the intricate canyonlands of the Río Balsas south of Tenochtitlán and eastward into the rugged mountain country of the Mixtecs. The more northerly route involved traveling east to the lowlands of the Gulf of Mexico, which were notoriously swampy and jungly, infested with mosquitoes, snakes, and other vermin, and crisscrossed by mighty rivers and splendid forests. A third route

5. We choose to refer to Precolumbian armed fighters as soldiers rather than the traditional "warriors." Modern scholars suggest that calling soldiers "warriors" portrays them in an imperialistic or paternalistic fashion, suggesting they were somehow more primitive or barbarian than fighters from European powers. See Hämäläinen 2022, xv.

6. Jiménez Moreno 1966, 23.

7. Fargher et al. 2011.

led to the Pacific coast, a strip so narrow, steep, and treacherous as to be impassable much of the year.

In this book, we present the more natural route, the two connected valleys and their complexities, their place in Mexico, and the features that have made them unlike any others in the world. The variety of biota, climates, cultures, landforms, and landscapes and their effects on history present a bewildering array of variables. We attempt to connect them in a way that accentuates their admirable diversity. A freeway now runs from near the sprawling city of Puebla to the city of Oaxaca. A more leisurely and instructive route is the same Mexico Highway 135 that we followed in our early visits, a winding, narrow roadway, where progress is slowed by hundreds of the notorious *topes*, speed-control bumps installed in every settlement along the highway to slow down speeding vehicles and (some cynics posit) to encourage travelers to patronize local business establishments.

We cannot discuss the valleys without invoking the peoples who have structured the human landscape over the last several millennia—Chochos, Cuicatecs, Ixcatecs, Mazatecs, Mixtecs, Nahuas (including one especially influential group, the Aztecs), Olmecs, Popolocas, and Zapotecs, to name the principal cultural groupings that have survived. They and their ancestors built the great early civilizations of Mesoamerica, including San Lorenzo of the Olmecs; Teopantecuanitlán, Guerrero, with its unknown founders; Teotihuacan, the megapolis of early Mexico; Monte Albán, the Oaxacan mountaintop city of the Zapotecs; Cholula of unknown founders; Tula of the warlike Toltecs; the splendid jungle cities of the Maya; and Tenochtitlán, the lake citadel of the Aztecs, civilizations that span three thousand years. All left traces in Mesoamerica, many of them lingering monuments. Representatives of all used the valleys as a transportation corridor or, in some cases, as home.

The Ramp

The city of Cholula, an international power a millennium ago, is now a virtual suburb of the great city of Puebla, which was founded by Spaniards. Cholula's location is ideal for beginning a reconnaissance of the valleys to the south. It lies at an elevation of roughly 2,150 meters (7,000 ft.) in the state of Puebla toward the southern end of the basin of Puebla and plateau of Mexico, an edge formed by the TMVB. The environs of Cholula, at least as they were prior to the mass urbanization of the late twentieth century, consist mostly of gently rolling country, covered in oaks, conifers, and tens of thousands of milpas (peasant cornfields).[8]

8. The milpa is the basic agricultural unit of Mexico. While principally a plot dedicated to corn, the milpa has nearly always been used for raising numerous other crops at the same time, including beans and squash,

FIGURE 1.1 Satellite view of the dormant volcano known as La Malinche or Matlalcueitl (blue skirt in Náhuatl), located near Cholula. Its summit lies at 4,420 meters (14,501 ft.) above sea level. Note the myriad fields, mostly cornfields, known in Mexico as milpas. Fields continue to encroach on the forests—the dark green areas that surround the summit. Courtesy Google Earth.

Fifty kilometers or so southeast, the plateau's edge tilts eastward to southeastward. At the small city of Tlacotepec de Díaz, the down ramp takes on the semblance of a valley, which continues for 130 kilometers (81 mi.) south.[9] Elevation along this ramp slowly drops, from over 2,100 meters (6,890 ft.) at Cholula to about 1,700 meters (5,577 ft.) at the city of Tehuacán, 150 kilometers (93 mi.) southeast of Cholula. As the ramp descends, rainfall decreases from nearly 1,000 millimeters annually at Cholula to around 500 millimeters at Tehuacán, to about 440 millimeters at the southern end of the valley, where the elevation is 515 meters (1,690 ft.). That low point records some of the hottest temperatures in the Atlantic Ocean slope of the states of Puebla and Oaxaca.

The city of Tehuacán is situated in this valley, now broad and fertile, part of the ramp that continues to tilt downward toward the south. Tehuacán has long been

---

all of which played a vital role in milpa production. The milpa continues to provide the economic foundation for much of rural Mexico, especially in the south.

9. For this study, we adopt the valley boundaries recommended by the Tehuacán Project. F. Johnson 1972, 8, figure 1. We also follow the project's tradition of including the Zapotitlán Valley in discussions of Tehuacán Valley.

FIGURE 1.2 Citlaltépetl, Pico de Orizaba, Veracruz, in 2003. At 5,636 meters (18,491 ft.) elevation, this dormant volcano is the highest point between Alaska and South America. Its companion is Sierra Negra, at 4,640 meters (15,223 ft.). In the last decade, snows have vastly dwindled, and the ancient glaciers have largely disappeared.

known for its *manantiales*, artesian wells and springs whose waters, now gravely diminished, since antiquity have purportedly been endowed with healing and life-prolonging properties. The life-giving waters originated in the ice and snow of Citlaltépetl, the great stratovolcano to the northeast. At 5,636 meters (18,491 ft.) above sea level, it is the highest mountain peak between Alaska and Colombia. The great peak, its snows now diminished, is satisfyingly visible from many points in the valleys.

The celebrated *manantiales* of Tehuacán made agriculture possible, for the valley is clearly more arid than the plateau above, and its fields seldom yield harvests without irrigation. A glance eastward explains why: the massive Sierra Zongolica, which rises 2,800 meters (9,186 ft.) to the east, traps moisture streaming in from the weather systems of the Gulf of Mexico and wrings that moisture from the clouds, leaving little for the valley. The Sierra Zongolica parallels the valley, forming its eastern flank, until to the south the range merges into the Sierra Mazateca, with summits exceeding 3,000 meters (9,843 ft.). Inside the state of Oaxaca, the lofty range continues, meeting there and merging into the Sierra Juárez (Sierra Norte, also known in this area as the Sierra Cuicatlán), which continues to flank the valley, now known as La Cañada.

FIGURE 1.3 Lower (southern) valley of Tehuacán, summertime. To the left of center is the village of San José Tilapa. Above center stretches the Sierra Mixteca.

At Tehuacán, the ramp drops sharply for a few kilometers, then continues more gently for another 50 kilometers until foothills and steep mountains on either side of the valley converge. Around Coxcatlán, site of the famous cave where archaeologists uncovered early fragments of domestic corn, the Río Comulco enters from the east, emptying into the Río Salado. The river that drains the valley assumes that name once the Río Zapotitlán merges from the west with the Río Tehuacán.

The valley of the Río Zapotitlán, located southwest of the city of Tehuacán, is often considered part of the Valley of Tehuacán. It is geographically distinct, however, even though it drains into the lower valley of Tehuacán. Indeed, its flora is even more varied and stranger than those of the larger and more prominent valley to the east. Twenty kilometers south of Coxcatlán, around the small town of Tecomavaca, Oaxaca, the mountains join, forcing the drainage into a canyon through which the river has cut a passage over the ages.

Seven kilometers south of Tecomavaca, the Río Salado meets the Río Grande. This much larger river flows northward out of a profound canyon, a breach in the Sierra Juárez, whose summits exceed 3,000 meters (9,843 ft.). The two rivers form the Río

FIGURE 1.4 View from Cerro Quiotepec of the merged rivers, the muddy Río Salado (*left*) and the larger Río Grande (*right*), that form the Río Santo Domingo.

Santo Domingo. The joined waters, the muddy Río Salado flowing alongside the clearer Río Grande, gradually mix, then plunge eastward through a deeply incised gorge that splits the two mountain ranges, eventually forming the Río Papaloapan in the coastal lowlands of Veracruz, a mighty river that discharges into the Gulf of Mexico.

If the Valley of Tehuacán can be viewed as a colossal ramp, the Valley of Cuicatlán and its tributaries to the south can be viewed as a staircase into the central valleys of Oaxaca. The valley bottom is often pinched into a narrow canyon by encroaching mountains only to open once again where side canyons have eroded the hillsides and deposited sediments sufficient to construct small tributary alluvial fans. These form the stairs of the staircase via a side drainage known as Cañada Chica and reach nearly to the valley of Etla, a few kilometers north of the city of Oaxaca. Through the main canyon flows the Río Grande, a river vastly larger than the puny Río Salado of the Valley of Tehuacán.

The Salado is an ephemeral river—often dry during the hot, rainless spring. It once flowed year-round, fed by myriad seeps and creeks from side canyons and valleys but mostly by the springs that also watered Tehuacán. Those once gushing

FIGURE 1.5 The Río Salado below Cerro Quiotepec and the breach of the Río Santo Domingo. The Río Grande is hidden on the right side.

manantiales are now reduced to a trickle, victims of excessive pumping and diversion. Even more damaging to the water supply has been the mistreatment of the great Citlaltépetl, source of the waters. Its forests are being ravaged by logging, and its once extensive glaciers and snowpack are melting, victims of global climate change.

Tehuacán has become a bustling, sprawling industrial urban center. Agriculture in Valley of Tehuacán has come to rely less on the *manantiales* and more on water derived from pumping or draining into the valley from the Sierra Zongolica to the east, the Sierra de Zapotitlán to the west, and, farther south, from the Sierra Mazateca to the east and the Sierra Mixteca to the west. Over the last several millennia, the valley's people have learned to collect water for their crops and needs by developing elaborate and highly managed systems of controlling runoff, as we shall see.

As if to underscore the extensive and elaborate system of irrigation, myriad milpas surround the small cities south of Tehuacán—Altepexi, Chilac, Miahuatlán, and Zinacatepec, towns where Náhuatl is often the lingua franca of the markets.

A journey down the ramp produces rapid and noticeable changes in vegetation. By the time one reaches Tehuacán and continues south and southeast, the oaks and pines of the highlands have given way to vegetation with a demeanor of aridity: low shrubs, scrubby trees, cacti, agaves, scrawny grasses, an assortment of yuccas and

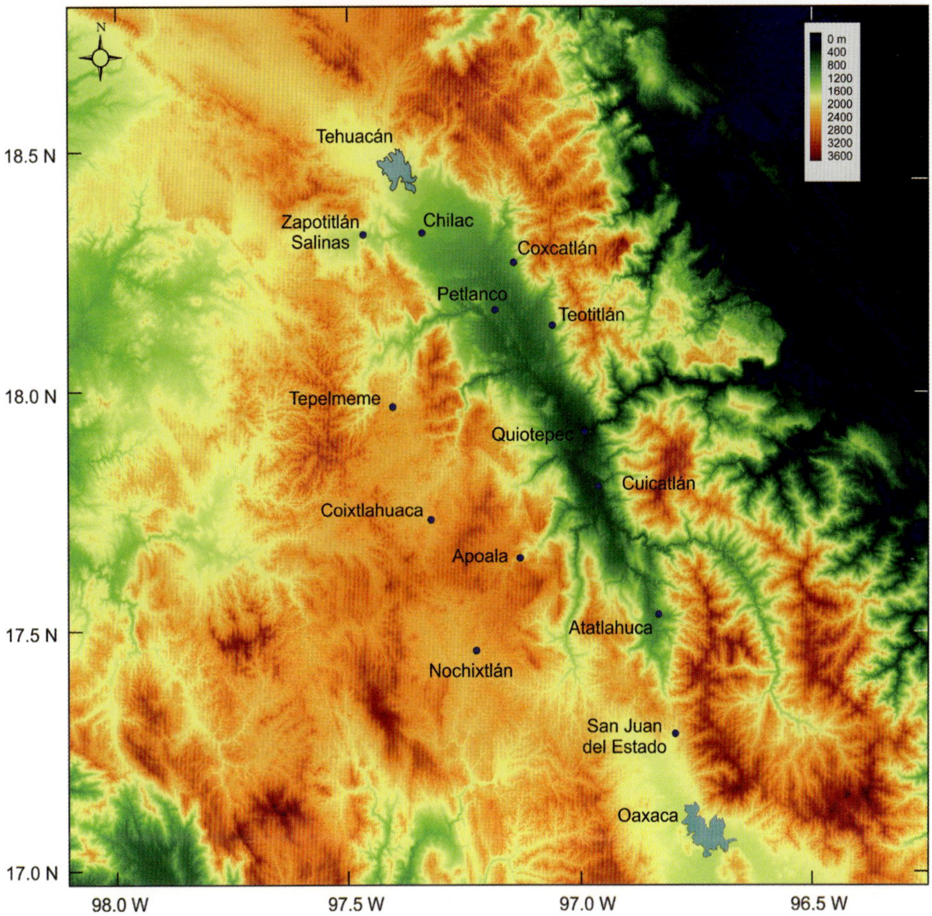

FIGURE 1.6 Digital elevation model of Tehuacán and Cuicatlán Valleys. The highest elevations are scarlet, while the lowlands of the Gulf coast are blue. The green, insect-like portion (insect facing downward) shows the valleys, with the Río Santo Domingo draining them to the right. The Valley of Zapotitlán enters from the west (*left*). The Río Grande (*far right*) appears as the insect's left antenna, and the Río de las Vueltas and Cañada Chica appear as the right antenna. The Río Tomellín (*lower*) and the Río Apoala (*upper*) appear as the insect's forelegs. The Sierra Zongolica is the mountain range at the top. The Sierra Mazateca appears just above the Río Santo Domingo. The Sierra Juárez extends east of the Río Grande, while the Sierra Monteflor appears between the Río Grande and the Río de las Vueltas. The valleys of Oaxaca are at the bottom right. From GMTED2010, U.S. Geological Survey.

yucca-like plants, as well as thorny acacias, mimosas, and mesquites. This sort of plant community is not unusual in Mexico, for much of the country is semiarid or arid. What distinguishes the landscape in these valleys from all others is the appearance of strange plants of otherworldly appearance. Massive columnar cacti, contorted yuccas of great height, and bushy trees with grotesquely swollen trunks dot

FIGURE 1.7 Tehuacán, Puebla, a city of nearly 250,000 residents. Once famous for its mineral waters, it now faces severe water shortages.

the landscape. South of the cities, forests of gigantic columnar cacti swathe each side of the valley in such densities that the hillsides take on the hue and bristly texture of the cacti. To the southeast, in the few remaining pockets of original vegetation, rich tropical deciduous forest, amply populated with tall cacti, represents a nearly impenetrable landscape. Toward the southern end of the valley, the side drainage of Zapotitlán enters from the west, home upstream to an even greater array of strange plant forms growing in abundance and in a variety seen nowhere else in the world. Grove after grove of giant cacti, several of them found only here, dot the landscape, along with other strange trees that attract plant explorers from the world over.

The natural history of the Tehuacán and Cuicatlán Valleys and their tributaries is intricate and varied. The unusual plants are so many and often so strange that Mexico has designated it as a Biosphere Reserve, and the UNESCO has followed suit.[10] The canyons and valleys are mostly arid or semiarid—deserts situated in the

10. The Mexican government designated it a Biosphere Reserve in 1999, and in 2018, the United Nations Educational, Scientific and Cultural Organization (UNESCO) designated it as a natural and cultural World Heritage Site ("originary habitat of Mesoamerica"). Theoretically, the designation by the Mexican government

FIGURE 1.8 Landscape near Texcala, Valley of Zapotitlán. The large yuccas are *izotes* (*Yucca periculosa*). The tall columnar cacti are *Mitrocereus fulviceps*.

tropics and objects of wonder to scientists and visitors from throughout the world for their climatic and microclimatic variations, their topographical and geological complexity, and, of course, the distinctive plants that grow and live there. The valleys' role in the domestication of crops—the valleys have some of Earth's earliest records of human uses of corn, avocados, and chiles—has gained international attention. Their place in the history of socially organized Mesoamerican peoples—for at least three thousand years—is equally significant, though less celebrated.

---

bestows some sort of protection for the valleys' natural values. Practice is sometimes a different matter: In 2004 we hiked through a small valley and hillside adjacent to a roadside near Tecomavaca. It was a place of spectacular density of cacti, nearly impenetrable, especially *Pachycereus hollianus*, but with impressive stands of *Pachycereus weberi*, *Neobuxbaumia tetetzo*, and *Myrtillocactus geometrizans* as well. We marveled at the density of the forest and its pristine condition. In about 2008 the valley and hillsides were bulldozed, destroying all the cacti. That irreplaceable habitat is now a plantation of lime and orange trees. In 2014, we viewed a large quarry for mining travertine that was excavated in the heart of the valleys, virtually decapitating a hill with a fine population of critically endangered *Fouquieria purpusii*, as well as a reportedly thriving population of *Dioon purpusii*. These examples demonstrate that designation as a Reserva de la Biósfera does not appear to impose constraints on development, mining, or agricultural expansion. Nonetheless, the designation probably has halted some major developments that otherwise would have been carried out. It also provided a sense of place, ownership, and pride to local inhabitants, promoting conservation and management of common resources in some communities.

CHAPTER 2

Deep History, Climate, and Hydrology or, Why the Valleys Are Deserts Within the Tropics

The transition from temperate forests and fertile plains of the plateau to hot desert in the valleys is dramatic. In fact, Tehuacán and Cuicatlán Valleys contain the southernmost deserts in North America. Yet they lie far from the subtropical horse latitudes, those areas between 30 and 35 degrees from the equator in both hemispheres, where most of Earth's deserts are found. The valleys' deserts lie entirely within the tropics and are bounded to the east and west by areas of luxuriant montane and tropical vegetation. This zone of unexpected aridity has given rise to the natural and cultural history that sets the valleys apart from the rest of North America and the world. To understand why this desert region appears in the tropics, we must look to the mountains and how the rain shadows they produce yield deserts. And to understand rain shadows, we must look to geography and thermodynamics.

First, we briefly describe the global atmospheric forces that influence the valleys' climate. Mesoamerica is sufficiently close to the equator to be directly affected by annual and seasonal variations in the Intertropical Convergence Zone (ITCZ), the great weather dynamo of Earth's equatorial regions. The changing position of the ITCZ through the year influences rainfall distribution and intensity, producing seasons defined by greater or lesser rainfall rather than by temperature, as is the case with the typical four seasons of temperate latitudes. The ITCZ results from global weather patterns that produce six belts of winds, three on each side of the equator: the Hadley cells in the equatorial latitudes, a circulation that produces the ITCZ; the Ferrell cells in the horse latitudes (around 30° latitude), where circulation of dry air from the upper Hadley cells descends, resulting in most of the world's extensive deserts; and the polar cells, a circulation that produces the polar front at about 60° latitude and high pressures in the poles, the origin of most wintry air masses in the

Northern Hemisphere. These six cells—as well as Earth's major ocean currents—are launched into motion by the rotation of Earth through a phenomenon known as the Coriolis effect combined with solar heating of the tropics. Subtle changes in the dynamics and trajectory of the ITCZ, such as El Niño Southern Oscillation, can result in droughts or flooding in nearby areas. Monsoons are a result of the seasonal migration of the ITCZ and the peculiar topography of the Sierra Madre Occidental of northwestern Mexico and the southwestern United States.

In addition to the effects of the changing position of the ITCZ, the desert climate of the Tehuacán and Cuicatlán Valleys is a result of the trade winds; the tectonic dynamics of southern Mexico, as seen in the complex topography of the Sierra Madre del Sur jutting against the TMVB (rain shadow); and, underlying all other factors, the physics of air masses. All combine to produce this peculiar arid and semiarid topographical complex, which in turn has given rise to the valleys' special vegetation, their hydrology, and the way of life of their peoples. This combination also attracted our attention: we were students of the Sonoran Desert. Our first exposure to the valleys produced impressions of both familiarity and wonder at the likeness to our desert homelands.

Trade winds were also a major factor in the European Conquest of Mesoamerica. These oceanic winds—semitropical easterlies blowing from North Africa to the Americas, and temperate westerlies blowing in the opposite direction, from the Americas to Europe—powered the sails of European ships and allowed the colonization of tropical America and the extraction of its riches to the benefit of European powers. The changing location of the ITCZ through the year determined the sailing times of the Spanish treasure fleet (and the pirates that preyed on it): it departed from Seville in southern Spain during March–April to take advantage of the trade winds blowing southwesterly and returned during June by way of the northeastern-blowing westerlies, setting out just before the hurricane season.[1] The ebb and flow of the dry and wet seasons were as predictable in pre-Hispanic times as today. But variations in those same weather patterns produced stormy seas and hurricanes that imposed a heavy levy—a sort of climatic vengeance—for the tons of ransacked gold, silver, and precious stones, booty that the conquerors loaded on their ships bound for Europe. With the rise in global temperatures, the pace of storms, their variation, and the energy in the atmosphere available to them has increased, which has led to greater variability in the ITCZ and, in turn, to droughts of greater severity and storms of greater intensity. The rate of shipwrecks, unnerving to investors in colonial times, would be far greater now. The slave trade and rum traffic—highly profitable enterprises that maintained international economies—would have slowed considerably.

1. For a discussion of the historical importance of trade winds and westerlies, see Crosby 2004, 104–32.

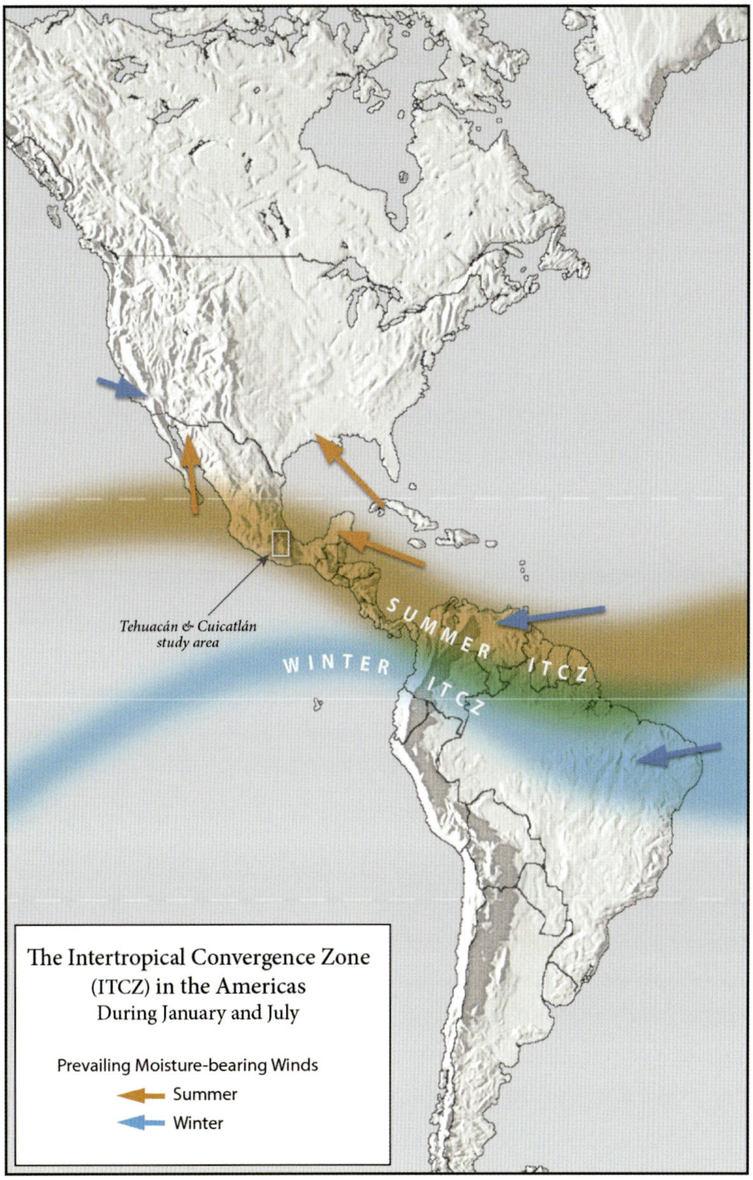

FIGURE 2.1 Diagram of summer and winter average position of the Intertropical Convergence Zone in the Americas, showing the prevalence of drier winters and rainy summers throughout Mexico. The arrows show prevalent summer and winter winds.

The Tehuacán and Cuicatlán Valleys owe their arid and semiarid climates to the rain shadow effect, the meteorological phenomenon that makes one side of a mountain range drier than the other. The mountain masses of the Sierra Zongolica to the east of the central and northern Valley of Tehuacán, the Sierra Mazateca to the east of the southern portion of the valley, and the Sierra Juárez east of the Valley of

Cuicatlán crest at altitudes ranging from 2,900 to 3,200 meters. This is sufficiently high to block the passage of rain-producing weather systems and cast a shadow of dryness on the lee side and on the valleys to the west. The trade winds blow warm and moist oceanic air overland—westward from the Gulf of Mexico—and propel these wet air masses toward the tops of the mountains, where they produce abundant rainfall, often heavy, but at other times, intermittent drizzle and fog for weeks, before crossing the highest ridges. The air masses continue their westward movement, the air now dry, having crossed the highest mountains. The dewatered air slides downward, not only producing clear weather, but also sucking any available humidity from the western slopes. Residents of the valleys for eons have watched longingly as dark clouds and thunderheads gathered along the mountain ridges to the east but denied the residents the pleasure of their moisture.

In thermodynamic terms, the simple formula $Q = T + W$ explains the whole process. The energy (Q)—heat content—of an air parcel moving in the atmosphere can be altered only by changes in temperature (T) or by producing work (W). As the air parcel rises, it expands as a result of lower pressure with increasing altitude. It cools in the process. When it descends, the weight of the atmosphere above compresses and thus warms it. If the change in the air parcel is adiabatic (no energy is added or removed), it will behave as a humid air parcel, changing its temperature by 6°C for every 1,000 meters in elevation. If energy (in the form of rainfall) is removed, it behaves as a dry air parcel, and the change in temperature will be 10°C every 1,000 meters.

If we were to follow an air parcel originating between 18° and 19° latitude north—the valleys' position—we would see warm air laden with humidity from the Atlantic and the Gulf of Mexico, carried by the trade winds over ocean waters for hundreds of kilometers, finally encountering land along the Mexican coast. Usually, the air parcel condenses rain even before it touches land, because the saturated air from the warm ocean cools and thus loses water while being pushed upward. The results of the landfall of this moisture-rich air mass are impressive. The lush tropical evergreen forests of Mexico's Gulf coast region are fed by more than 2,000 millimeters (79 in.) of annual precipitation along the coastal plain. The windward side of some steep mountain areas immediately adjacent to the Gulf, such as the Sierra Santa Martha in Los Tuxtlas region of southeastern Veracruz, may receive more than 4,000 millimeters (157 in.) of water annually. The air parcels travel unimpeded across roughly 250 kilometers of the lowlands of Mexico's Gulf of Mexico region, yielding a generous share of precipitation, sometimes resulting in too much rainfall.[2]

2. To the frustration of Mexican water authorities. The abundance or excess of water in the Gulf lowlands and to a lesser extent in the Pacific coasts in the south contrasts with the deficiency of water in the cooler highlands, especially the TMVB, where the bulk of Mexico's population resides.

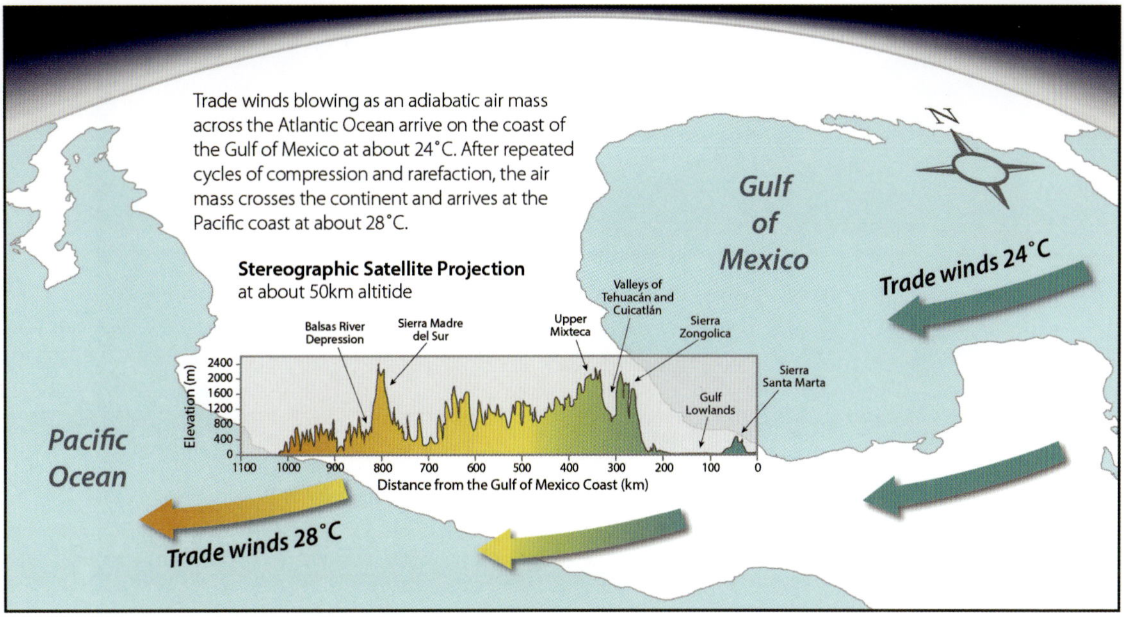

FIGURE 2.2 Diagram of trade winds as they affect the valleys and the mountains of central Mexico.

In these lowlands of once-towering tropical forests, the Olmec people flourished, producing one of the first great civilizations of Mesoamerica, beginning about 3,500 years before present. West of the Olmecs rise the summits of the Sierra Zongolica; the northeastern limits of the Sierra Madre del Sur stand as a tall fortification, rising to 2,500 meters; while the highest ridges of the Sierra Mazateca and the Sierra Juárez, south to southeast-trending extensions of the Sierra Zongolica, both exceed 3,000 meters. These ranges form a spur of the Sierra Madre del Sur that begins at the Isthmus of Tehuantepec, where it trends northwest and continues unbroken until it encounters the much younger TMVB in the north. Its eastern slopes present a lengthy sponge, absorbing Gulf moisture and hoarding it, releasing it for the benefit of lower elevations. Parts of the Sierra Zongolica's eastern slopes have proved to be ideal coffee-growing regions.[3]

Pressed by the prevailing wind, the warm and moist air parcels from the Gulf are forced upward when they encounter the steep mountain slopes. There they cool rapidly, losing humidity along the route and producing magnificent, antediluvian-looking vine and epiphyte-rich cloud forests. The high ranges not only slow the winds, but also produce cool and wet environments—numerous rivers, moist ravines, and dense, jungly vegetation, with abundant trail-choking vines and thick

3. Rodríguez 2003, 22.

FIGURE 2.3 Mesophytic cloud forest, east of the summit of Sierra Juárez. Rainfall here exceeds 2,500 millimeters (98 in.) annually. Elevation ca. 2,000 meters.

forests, which form a barrier to the passage of humans and their culture from east to west. Only the canyon forged by the Río Santo Domingo, which combines the waters of the Río Grande and the Río Salado, breaches the great barrier between the Isthmus of Tehuantepec and the Central Basin of Mexico and provides a narrow corridor of moist air to reach Quiotepec, Oaxaca, and its immediate vicinity.

For most of the westward journey, the air parcels behave as a wet adiabatic system. Their temperature when touching land at the shores of the Gulf of Mexico during the peak of summertime is usually a balmy 25°C (77°F). Higher up the mountains, the temperature lowers significantly. For example, at Fortín de las Flores, Veracruz, at 1,000 meters, the mean temperature of the air parcel is 20.9°C; in Acultzingo at 1,700 meters, it is 17.6°C. At the town of Zoquitlán in the Sierra Zongolica at 2,077 meters, the mean temperature is 15.2°C, and annual rainfall averages 1,160 millimeters (46 in). The highest peaks of Sierra Zongolica, at 2,350 meters, have an annual mean temperature of 13°C (55°F). At that latitude the air parcel has crossed mountains that average about 2,200 meters elevation. In the Sierra Mazateca, at the Mazatec town of Huautla, elevation 1,678 meters, where María Sabina offered psilocybin mushrooms to the world, mean air temperature is about 17.2°C, with annual rainfall of around 2,500 millimeters (98 in.). In Huautla's convoluted topography, the lush forests and persistent fog on the windward side attest to the pervasive humidity. One can look down to the east to a nearly perpetual fog bank (see figure 7.57).

At about 280 kilometers from the coast, the high ridges of the sierras drop precipitously into the Valley of Tehuacán. From the summit one can verify the effects of the first law of thermodynamics, for the valley below to the west is usually visible, that is, free of fog, and (except for irrigated fields) presents a parched appearance, contrasting sharply with the lush, verdant mountain forests and lowlands to the east. As the air parcel descends into the valley, it no longer behaves as a wet adiabatic. The more it goes downhill, the more it dries, compresses, and heats. As the system can no longer produce rainfall, it warms faster than a wet adiabatic for a given change in elevation.

The rainfall and temperature differences within the valleys themselves are striking. Toward the northern end of the ramp in Tehuacán, temperatures are milder than farther south because the air cannot descend any farther, while at Quiotepec and Cuicatlán, at the very bottom of the valleys, annual temperatures are nearly always warmer than those in similar elevations on the windward side of the Sierra Zongolica. As the valley ramps down from Tehuacán toward La Cañada, the increase in temperatures from north to south is a product of the lower elevation and the change of the air parcel into a dry adiabatic mass. At Tehuacán, elevation 1,700 meters, the average annual rainfall is 472 millimeters, while the mean annual temperature is about 18°C. At the town of Coxcatlán, the famous site of MacNeish excavations and the discovery of the Coxcatlán Corn Cave. at 1,145 meters elevation, not yet in the valley bottom, annual mean air temperature is 24°C, and the precipitation is 439 millimeters. Most of the moisture falls during the summer rainy season, derived from the Pacific Ocean. In Quiotepec, at the very bottom of the valleys, elevation 543 meters, or 1,200 meters lower than Tehuacán, mean temperature is 25.5°C, averaging 7.5 degrees higher than that of Tehuacán.[4] Not only are the air parcels drier at the lower elevations, but their increased heat produces more rapid evaporation and transpiration, resulting in arid conditions.

As we continue to follow the air parcel westward, we find the terrain climbing again, more gently, but still sucking humidity along most of its path, except in the highest mountains. The mass continues to Zapotitlán Salinas and climbs upward

4. Annual rainfall at Quiotepec is 477 millimeters, nearly equal with that of Tehuacán. Quiotepec derives additional rainfall from its mountainous location and, in all probability, its location at the upper limit of a canyon that breaches the mountains and gives Quiotepec direct access to the coast with its moist air masses. Axuxco, Puebla, some 40 kilometers distant and considerably higher at 950 meters elevation, is situated mostly in a broad plain with only low hills in the vicinity. Rainfall there is a scant 360 millimeters.

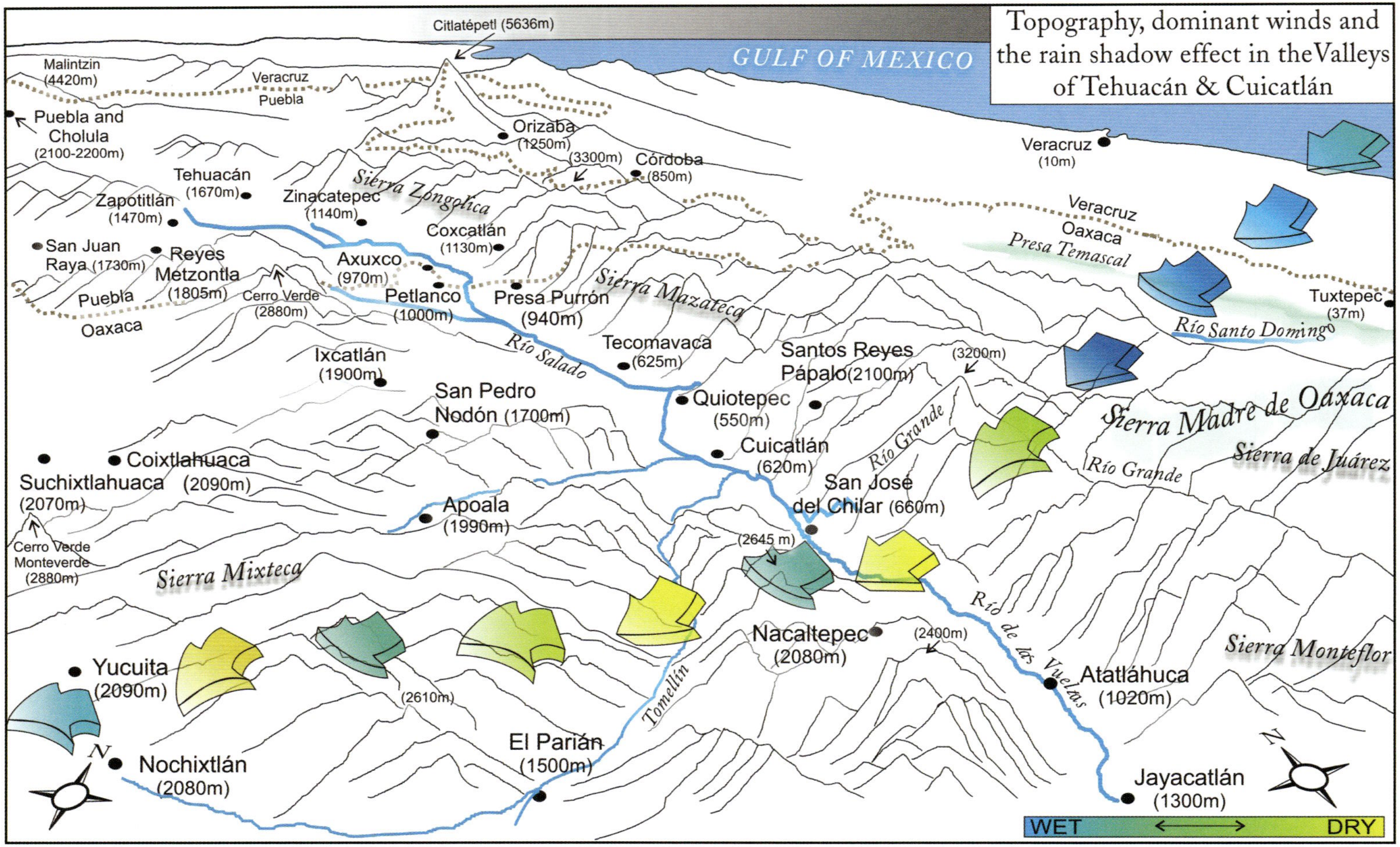

FIGURE 2.4 Diagram of dominant wind pattern over the Gulf of Mexico coast and the Tehuacán and Cuicatlán Valleys. Arrows show a single air parcel entering land as brisk, humid winds releasing rainfall in the tropical wet forests of the Gulf and the east-facing high sierras. The rain shadow effect happens as soon as these winds descend into the valleys on their westward course across central Mexico, where they become warmer, dry air masses (yellow arrows).

FIGURE 2.5 Zapotitlán Valley from near Acatepec, looking northeast toward the Sierra Zapotitlán. Elevation ca. 2,000 meters.

toward the Sierra Mixteca, a comparatively dry complex system of highland mountains, where the Mixtec culture developed. At this point—around 2,000 meters, at the town of Acatepec—it crosses the Continental Divide, leaving the Río Papaloapan basin, which drains into the Gulf of Mexico. Acatepec, at 2,053 meters elevation, averages 16.4°C annual temperature but receives only 555 millimeters of rainfall, compared with Zoquitlán at the same elevation in the Sierra Zongolica, which is somewhat warmer but receives more than twice as much rainfall.

The air parcel now enters the extensive Pacific drainage depression of the Balsas River, where seasonally dry and torridly hot and dry tropical lowlands occupy the deeply incised canyons. By virtue of the cycles of rarefication (decompression) and compression in its up and down sojourn since encountering the coast at Veracruz, the air parcel is now much warmer, so much warmer that at the coast of the Pacific Ocean, once again at sea level, the air parcel is 10°C (18°F) warmer during the summer peak than at its starting point in the Gulf of Mexico. The simple physics of moving adiabatic air parcels explains the mild, wet conditions in the Gulf of Mexico

FIGURE 2.6 Landscape at the upper end of Zapotitlán Valley near Acatepec, a sui generis association of chaparral and thornscrub, including at least four *Agave* species plus Burseras, barrel cacti, and several deciduous and evergreen shrubs.

FIGURE 2.7 Chaparral with *Mitrocereus fulviceps*, dry season (December), in Cañada Chica. Flowering stalks are those of *Beaucarnea stricta*.

FIGURE 2.8 Same scene as in figure 2.7, but in the rainy season (July). Note that the *Beaucarnea stricta* are barely visible.

basin, and the intensely hot and dry weather of the Pacific coast. It also explains much of the arid climate of the Tehuacán and Cuicatlán Valleys. It is common in the winter months to find the valleys sunny and dry while the eastern slopes are enveloped in persistent clouds, fog, and drizzle.

The movement of Atlantic air masses explains only part of the climatic story of the valleys and their arid climate, however. The Intertropical Convergence Zone affects the Pacific region of Mexico as well as the Atlantic. In the winter of the Northern Hemisphere, the ITCZ concentrates its activity mostly south of the equator, where heating is greater, and the Pacific coast and coastal mountains remain mostly dry. The dominant vegetation in the coastal area is dry tropical forest (tropical deciduous forest), which is dormant during the winter, reflecting the lack of winter rains due to the drying out of air masses originating in the Atlantic. As summer approaches in Mexico's tropics, the Pacific Ocean and the equatorial landmasses heat up, and the atmospheric circulation from the ITCZ slowly moves northward, following the path of the rise in heat. In late May and early June, thunderstorms from moisture derived from the Pacific begin to develop over southern Mexico. These often-brief showers are heaviest along the coast and on the western (windward) slopes of the Sierra Madre del Sur for the same adiabatic reasons that precipitation derived from

Atlantic or Gulf of Mexico air masses is heaviest on the Atlantic side of the Sierra Zongolica. As the influence of the ITCZ pushes north, the storms intensify, spilling over into the valleys and reaching the coastal lowlands of the Gulf of Mexico as well. The Sierra Mixteca, which lies west of the valleys, intercepts some of the moisture, but these slopes are gentler, and peaks are for the most part lower than those on the Atlantic side, which thus permits more moisture to reach the valleys than is possible with moisture from the Atlantic. Hence the bulk of the rainfall in the valleys, stingy as it may be, falls between late May and mid-September and is deposited by afternoon thunderstorms. Still, the valleys are dry, especially compared with the cool, moist uplands of the pine-forested Mixteca Alta and the jungles of the Gulf coast. For seven to eight months of the year, the lower valleys' landscape illustrates the brown of drought.

But there are further complications to the story. The Pacific Ocean in the region of the Gulf of Tehuantepec in southern Oaxaca is also the nursery for tropical hurricanes. These potentially massive storms begin forming with the arrival of the advancing ITCZ and by mid-June initiate the annual hurricane season for southern Mexico's Pacific coast. The development and trajectory of these sometimes-powerful storms is erratic and is often confined to the coastal areas and those immediately inland. The earlier storms typically head west into the Pacific, but most years also see rainfall, sometimes intense, associated with the arms of the hurricanes over inland Oaxaca and extreme southwestern Puebla, especially late in the season. As our planet warms, these storms appear to be larger and more intense, though not necessarily more frequent. A hurricane of 2011 deluged La Cañada with unprecedented amounts of rain, opening hundreds of new and highly visible erosion scars on the hillsides and scouring and permanently widening the floodways of the Río Grande and the Río de las Vueltas.

Hurricanes that originate in the Atlantic or the Gulf of Mexico may affect the valleys as well. Atlantic hurricanes usually assume paths that take them north of the valleys, however, so they are less likely to be inundated by rainfall originating from Atlantic hurricanes than by those originating from the Pacific storms. Hurricanes originating in the Gulf of Mexico may stall for long periods over Central America and often hit the Yucatán Peninsula and travel westward across it. Their effects on the valleys are generally indirect; that is, they are felt as an increasing number and intensity of thunderstorms rather than storms with high winds.

To complicate matters even more, Mexico lies along the path of cold air masses originating in northern Canada, which spread as frontal systems south and east over northern Mexico and the Gulf of Mexico during the winter months. Since no significant mountain ranges impede their southerly progress, they eventually descend into the tropics. Given their northern origin and the direction of their accompanying winds, these are commonly referred to as *nortes* or, if blowing over the Isthmus and the Gulf of Tehuantepec, Tehuantepecers.[5] The frequency of nortes varies from year to year, with an average of about forty-four, but with as many as seventy or as few as twenty in a given year. These vary in intensity from powerful to barely discernible. Their effects last between two to six days. They mainly affect northern, central, and Gulf coast Mexican states, as well as the United States, including most of Texas. It is not uncommon, however, for these cold air masses to bring copious rainfall and low temperatures to most of eastern Mexico and to the U.S. southern border. In the Gulf coast of Veracruz, temperatures can drop up to 20°C (36°F) in a few hours, and sometimes only in minutes when cold fronts are fast moving. Rainfall can be up to 200 millimeters (8 in.) in a single day. At the southern edge of the northers, winds in the isthmus can attain sustained speeds up to 30–70 kilometers per hour (19–43 mph), with gusts of more than 100 kilometers per hour (62 mph), because of the large atmospheric pressure differential between the Gulf of Mexico and the Pacific. Nortes strongly affect Tehuacán and Cuicatlán Valleys and are responsible for over 90 percent of the winter rains and frosts in the Mixteca Alta. These regimes may be a factor in the persistence of chaparral in the valleys. This vegetation type flourishes under cooler temperatures and winter rainfall and is usually associated with Mediterranean-type climates.

This climatic portrayal describes the prevailing climates during the times of human occupation. Recent studies based on fossilized pollen derived from ancient geological strata suggest that the climate has been stable in the valleys since the Middle Miocene, more than 15 million years ago, though moisture was probably more abundant during the Pleistocene (1.8 million–12,000 years ago) than at present. Scientists suggest that this uniformity of climate has enabled ancient flora to persist in the valleys, while new species were evolving in the arid and semiarid climates, thus offering ideal conditions for both widespread endemism and plant diversity.[6] The summer rains and winter droughts, though subject to occasional variability, have thus been reliable for eons. The climates of the valleys guarantee to their flora a stable way of life—warm, arid to semiarid, and free of frequent massive storms—the ideal location for incubation and evolution of new species. Rainfall may be stingy, but it has been more or less dependable for a long, long time, perfect conditions for plants of the deserts and semideserts.

5. Magaña et al. 2003. Willis E. Hurd coined the term Tehuantepecers in 1929; Trasviña et al. 1995.

6. Ramírez-Arriaga et al. 2014, 25.

CHAPTER 3

Geology of the Valleys

Despite the desert conditions in the Tehuacán and Cuicatlán Valleys, people have been consuming plants and animals there for at least eight thousand years and have cultivated plants (and probably domesticated a few animals) for at least six thousand years.[1] This long uninterrupted history of human occupation, much of it characterized by settlements, is mostly a result of the valleys' relative geological and climatic stability over deep time, the last 10–20 million years. Although the valleys have experienced gradual cooling since the Middle Miocene, the flora is a complex and diverse set of taxa adapted to arid and semiarid conditions closely related to the extant flora of the region.[2] The stability of a deeply dissected desert surrounded by lofty mountain ridges has provided the perfect incubator for plant endemism—plants evolved in desert conditions and finding emigration from the valleys not at all to their liking. And the results are dramatic: of the 3,500 plant species cataloged so far, more than 10 percent are endemic.

Yet geological stability did not entail lack of dynamisms. During most of those millions of years, the valleys formed a basin, that is, a depression some 130 kilometers long and 5–15 kilometers wide, with no outlet to the sea. Rivers emptied into the basin and formed a lake between the high mountain ridges that hemmed it in. During Cenozoic time, water and wind transported sediments derived from the surrounding land into the basin. These would one day provide fertile soils for newly evolving crops, as early agronomists undertook genetic modification of wild plants, that is, began to domesticate them.

1. MacNeish 1967, 292, 305, fig. 188.

2. Ramírez-Arriaga et al. 2014.

FIGURE 3.1 Cerro Quiotepec and the Sierra Mazateca. The cerro, home to an archaeological site, is the hill with cliffs to the left of the village, Santiago Quiotepec. The canyon of the Río Santo Domingo appears just above the bottom center of the photo. It is formed by the junction of the Río Grande, which drains the Valley of Cuicatlán, and the Río Salado, which drains the Valley of Tehuacán.

For millions of years the lake level rose and fell as rainfall and climate varied, the rate of sediment accumulation and the nature of the sediments varying as well. But during the Quaternary, that is, during the last 2.5 million years, long before the arrival of humans, the ancient body of water was captured—overtaken and drained by the upstream cutting action of the Río Papaloapan of the tropical Gulf coast, which crept westward, attacking the mountains, gnawing its way into the growing channels of eastward-draining watercourses until it broke through the divide, gradually becoming the modern Río Santo Domingo as it chewed southwest into mountains.[3]

From an aerial view, the basin stands out, still hemmed in by mountains. But to understand how the surrounding mountains came to be and how the basin originated, we must travel back into geological times. The lake owed its existence to the peculiar configuration of the Sierra Madre del Sur, which in turn has resulted

3. Brunet 1967, 74.

from the geological evolution along the margin of the circum-Pacific belt in southern Mexico. This margin has long been the scene of the battle—collision and separation—of lithospheric plates, that is, colossal sections of Earth's rigid crust, in this case the Cocos Plate, which interacts with the North American and Caribbean Plates.[4] During tectonic extension, that is, plates pulling apart rather than colliding, the continental blocks stretched, producing a depression between separating blocks as the bottom dropped out of the intervening space.

To describe the geological situation with impossible brevity, the mountains of Oaxaca and southern Puebla—the Mixteca Alta and the Sierras Zongolica, Mazateca, and Juárez (also called Sierra Cuicatlán)—have risen primarily as a result of the deformation of rocks in the region during the Laramide orogeny. This was a geologically traumatic period of tectonic activity that gave rise to the Rocky Mountains in the United States and ancestral mountains in southwestern Mexico 70 to 40 MYA (million years ago). Much later, crustal melting during the subduction of the Cocos Plate in the Pacific Ocean under the North American Plate produced the Trans-Mexican Volcanic Belt (TMVB). The TMVB forms the east–west-trending string of volcanoes, extending more than 900 kilometers from near the Pacific Ocean, in the states of Nayarit and Jalisco, to Citlaltépetl (Mt. Orizaba), in western Veracruz. This procession of volcanoes forms the northern boundary of our area. As noted, the TMVB is also Mexico's most densely populated area. The complex movements of Earth's crust as the plates collided, subducted, and tore apart produced the variety of tilted uplifts and collapses as well as the volcanic activity that created the mountains of the TMVB.

This simplistic description of coarse structure labels the rough topography but omits the myriad separate variables and geological episodes that account for the topography as we see it.[5] Suffice it to say that geologists are still working out the details of the valleys' origins. In general, though, they agree that the mountains that surround the valleys have arisen because of three general geological processes. First, those forces have dehydrated the subduction plate (the Cocos Plate) and partially fused the upper plate mantle (the North American Plate), resulting in volcanic activity. This, in turn has given rise to the east–west-trending TMVB, which intersects and limits the northern end of the Valley of Tehuacán. As if to underscore the TMVB's terrestrial significance, the stratovolcano Citlaltépetl, Mexico's highest point, located at the TMVB's eastern edge, is visible from many of the valleys'

4. Geological interpretations of the assembling of contemporary Mexico are in a state of ongoing refinement. To quote geologists referring to the Oaxaca Fault region, "The morphological evolution of the basins in the Sierra Madre del Sur (SMS), southern México is poorly understood" (Medina-Sánchez et al. 2020, 6). We present only a rough description and acknowledge that details are subject to revision and refinement.

5. Keppie 2004, 767.

hilltops on a clear day. It is the third highest mountain of North America, after Denali and Mount Logan, located almost 6,000 kilometers northwest.

Second, mountains also rise from a succession of earthquakes, as continents jam, lithospheric blocks collide and fuse, or tectonic plates pass and grind against each other or pull apart. Earthquakes are a source of mountain building, or, better, mountain building is often the tangible fruit of earthquakes. Individual earthquakes may result in only minor reshaping of the crust, say, a jump (vertical or horizontal) of 3 meters along one side of a fault relative to the other. Repeated quakes, however, are thoroughly capable of producing towering mountains, such as the Oaxaca's highest peak, Cerro Nube Flan, in the coastal range of the Sierra Madre del Sur, which reaches an elevation of 3,700 meters. If one earthquake occurs only once every 150 years, in less than 200,000 years, a mountain of that elevation may result.[6] Geological forces always have ample time to carry out their results: mountain building, erosion, or collapse.

As the local inhabitants can attest, earthquakes are ubiquitous throughout the Cuicatlán-Tehuacán region: on June 15, 1999, an earthquake with a magnitude of 7.0 on the Richter scale shook much of northern Oaxaca and southern Puebla and destroyed or damaged more than twenty thousand homes. The epicenter lay near the town of Zapotitlán Salinas in the Valley of Zapotitlán, but major damage was inflicted on the city of Tehuacán, some 30 kilometers distant. In our early visits to Tehuacán, some buildings from the city's older downtown section remained closed because of earthquake damage.

On September 8, 2017, an earthquake measuring 8.1 on the Richter scale shook Oaxaca and Chiapas, causing major damage. Although it originated some 70 kilometers deep in the Gulf of Tehuantepec, it produced major shocks and created great alarm among the residents of Mexico City, more than 1,000 kilometers distant. It also produced massive destruction in the city of Juchitán in the Isthmus of Tehuantepec.

Most of the earthquakes in the region—and they are numerous and often lethal—result from that same subduction of the Cocos Plate, which is moving northward at about 75 millimeters per year and plunging beneath the North American Plate.[7] Small earthquakes are a regular occurrence throughout mountainous Mesoamerica, and major quakes are a constant threat. It is with good reason that anthropologist Eric Wolf referred to Mesoamerica as "The Shaking Earth."[8] Earthquakes occur

---

6. In roughly 5 million years, earthquakes along the northeast side of the "bend" in the San Andreas Fault have produced the uplifting of the San Bernardino Mountains in southern California to more than 3,300 meters (10,827 ft.).

7. Sedlock, Ortega-Gutiérrez, and Speed 1993, 5.

8. Wolf 1959.

when stress has accumulated to the elastic breaking point: something must give way. In each earthquake, one side of a fault shifts position relative to the other—up, down, or sideways.

Third, mountains may also rise based on the physics of individual terranes. Instead of a direct collision, plates may move sideways to each other or pull apart. These motions produce extensional stretching, during which the crust is torn apart at fault lines. As lithospheric blocks, floating on the Earth's mantle, are "liberated" from their former crustal block, they seek their center of gravity much as ice cubes floating in a glass of water rotate and sway when agitated before coming to rest. Some blocks rise, forming horsts. Neighboring blocks subside, forming grabens, as the tectonic motion unceasingly subjects them to the forces of gravity.

Stretching, as well as compression, is taking place in northern Mesoamerica because the subduction of the Cocos Plate under the North American Plate is not a simple dive-under-and-lift function. While it is subducting under the North American Plate, the Cocos Plate is also subducting under the Caribbean Plate, which is moving eastward with respect to the North American and the South American Plates. Much of southern Mexico, including Oaxaca and Puebla, is slowly moving southeast, parallel—not perpendicular—to the trench where the Cocos Plate is subducting under the North America Plate.[9] The three plates influence each other's motion. They meet at a triple junction in the Gulf of Tehuantepec in the Pacific Ocean. Such meeting places usually are the sources of considerable tectonic activity and were the source of the 2017 earthquake. And various physical forces they exert on the tectonic masses involved make for contorted landscapes.

Geologists are now able to trace this churning of land masses to its origin over the last billion years or so, during which time large blocks of crust have faulted and broken off, twisted, and rotated, suturing onto or tearing away from other terranes, becoming major forces in the interplay of the colossal drama of plates in motion. Interaction among the plates has produced a convoluted mountain mass in Oaxaca and Puebla, as well as a string of volcanoes to the east in Guatemala and in the Mexican state of Chiapas.[10] And they have given us the TMVB. To complicate the geological story even further, part of the Mesoamerican landmass consists of accreted terranes, chunks of former continents or island arcs, scraped or broken off from huge plates, as though by a ponderous razor or sledgehammer, and ultimately stitched, like a bold rag quilt, onto other large landmasses as they ground their way underneath, on top of, or affixed to adjacent plates or terranes. Or they

9. Sedlock, Ortega-Gutiérrez, and Speed 1993, 5.

10. See, for example, Alaniz-Álvarez and Nieto-Samaniego 2007, 85–86; El Chichón volcano in Chiapas erupted in 1982, causing around two thousand deaths, immense destruction, and disruption of agriculture. It produced unusually large volumes of sulfur dioxide gas.

may represent huge blocks of volcanic rock generated at the great oceanic rift zones (locations where plates are separating) and ferried onto continents as plates move around and jostle for position.

Perhaps the most readily visible geological marker of the valleys is the Oaxaca Fault. The events that produced the region's currently visible topography are relatively recent in the geological time scale, though the building blocks—and the visible rocks—may be far older. Geologists trace the formation of Mesoamerica to more than one billion years ago, but the infancy of the valleys dates only from the Late Cretaceous period (if roughly 80 million years could be a measure of small-time magnitude). Their single most notable structure is the Oaxaca Fault (or Oaxaca Shear Zone), which forms the eastern edge of both valleys. The fault extends northward from the Gulf of Tehuantepec southeast of Oaxaca City, through the District of Etla a few kilometers north of Oaxaca City in central Oaxaca, and continues as a mountainous escarpment, forming the eastern side of the valleys until it dwindles a short distance north of the city of Tehuacán.[11] This massive—and visible—geological feature has roots stemming from as long ago as the Jurassic period, more than 200 million years ago, well before the valleys were formed.

Understanding the origins of the Oaxaca Fault takes us back to the end of the Carboniferous period, roughly 300 MYA, during the Paleozoic era, the age of fishes. At that time all the Earth's continents formed a single landmass, a supercontinent called Pangaea, which floated on a single gigantic ocean called Panthalassa in what is today the Southern Hemisphere. The single mass and the southern position of Pangaea created a world very different from today's, with not only ancient and strange organisms, but completely different climates, wind patterns, and ocean currents.

During the Late Triassic (ca. 220 MYA), large portions of Pangaea began to tear away from the great mass, creating separate continents. The gaps—voids—between the original Pangaea and the departing lands developed into the world's seas. One ocean body that began to open was the early Gulf of Mexico. As North America and South America parted, huge blocks of continent stretched beyond the ability of the land to hold together, and water rushed in to fill the gathering void. Rather like the cheese on a hot pizza that collapses to the platter when a piece is pulled away from the rest of the pie, the ripping of the Earth's surface initiated the formation of long grabens–depressed blocks of land—that dropped into the developing empty space left by the tearing apart. The grabens, or sunken landmasses, in southern Mexico were bordered by parallel faults, with orientations from northwest to

11. The fault's movement north of Oaxaca City is acknowledged by a road sign near the town of Guelatao, the birthplace of Benito Juárez. The sign reads: "Precaución: Falla Geológica" (Caution: Geological Fault). The fault's movement causes ongoing cracking and buckling in the pavement near the sign. Precisely what precautions the traveler is urged to take are not specified.

FIGURE 3.2 Cuicatlán, the Río Grande, and the Oaxaca Fault seen from near Jocotepec, Mixteca Alta.

FIGURE 3.3 Cliffs of the Oaxaca Fault zone, near Cuicatlán.

southeast. Tectonic processes—collisions, rifting, shearing, thrusting, twisting, and displacing—gradually gave shape to our valleys. By the Late Cretaceous, after roughly 80 million years, the Sierra Madre Oriental, including Sierra Zongolica and the Sierra of Chiapas toward the southeast, started taking shape. During the Cenozoic, beginning about 65 MYA, the region finally acquired its present structure and morphology, and the crustal blocks that shaped the valleys were more or less in place.[12] The already ancient Oaxaca Fault zone was even then featured in prominent relief.

The Oaxaca Fault itself became a major driver of change in the region's gross topography. Though its origins date back at least 200 million years, the fault shows up on the landscape in various phases, from Laramide events (the same tectonic forces that built the Rocky Mountains) during the Late Cretaceous and Paleogene (70–40 MYA) to the final large faulting events during the Miocene, a process involving a leisurely 30 million years or so. Only a very slowly recording time-lapse camera could reveal the drama of the fault's evolution.

Geologists can identify crustal or terrane boundaries, which are major faults or basins separating adjacent blocks with different geological history and distinct rocks. The block that forms the eastern side of the Oaxaca Fault is often known to geologists as the Cuicateco, or Juárez, terrane and comprises much of the sierras at the valley's eastern edge. The fault's reddish-pink cliffs form sections of the western side of the block (the eastern side of the fault) and can be seen from a great distance in the vicinity of Cuicatlán north to Tilapa, Puebla. Indeed, if we stand on the western side of the Río Grande at Cuicatlán and look eastward at sunset, the deep rose cliffs stand as a tribute to geological processes hundreds of millions of years in the making. East of Tehuacán as well, the fault is exposed, locally referred to as Cerro Colorado.

West of the Oaxaca Fault lies the Zapoteco, or Oaxaca, terrane, another crustal block that is host to our valleys, their ground floor.[13] It is a former oceanic island arc that has sutured onto the North American Plate as the plates collided.[14] The Zapoteco terrane has now dropped between the two adjacent blocks, forming a graben (or, according to some geologists, a half graben), as they have stretched apart, hence the gradual appearance of the valleys that came to form a huge basin.[15] The oblique fault segments, or echelons, of the Zapotec terrane are named (from north to south) most appropriately: Tehuacán, Coxcatlán, Teotitlán, Dominguillo, Jayacatlán, and Etla, beginning just north of Tehuacán and ending just north of Oaxaca City.

12. Dávalos-Álvarez et al. 2007.

13. Later published studies label a larger terrane as Oaxaquia. Campos-Enríquez et al. 2013, 62–73.

14. Geologists frequently introduce delightful and creative variations of more familiar terminology that become eminently descriptive metaphors.

15. Ramírez-Arriaga et al. 2014: 4; Dávalos-Álvarez et al. 2007.

FIGURE 3.4 Oaxaca Fault zone and sugarcane fields near Coxcatlán, Puebla. Here, the fault is a much lighter color than it is farther south. (See also figure 6.22.)

The mountains that lie generally west of the Zapoteco terrane (and the valleys) constitute the Mixteco terrane. These mountains include most of the Mixteca Alta and are separated geologically from the Zapoteco terrane by the Caltepec Fault, named for the village of Caltepec, which lies some 10 kilometers south of the Popolocan town of Los Reyes Metzontla.[16] Between the Oaxaca Fault and the Caltepec Fault lie the Valleys of Tehuacán and Cuicatlán, constituting a block known to geologists as the Tehuacán graben, part of the Zapoteco terrane. The Caltepec Fault running parallel to the Oaxaca Fault is interpreted as part of a tectonic system that recapitulates the deep history of the valleys: as the valleys drop between the bordering horsts, they have been releasing tectonic stresses associated originally with the rupture of Pangaea, with periods of tectonic reactivation from the Cretaceous until the present, and periods of tectonic quiescence during Triassic, Jurassic, and Middle Cretaceous times.[17] Early in the Cenozoic (65–2.6 MYA) the terranes began to move apart at the fault zones as the Caribbean Plate began to migrate to the east,

16. Sedlock, Ortega-Gutiérrez, and Speed 1993.

17. Elías-Herrera et al. 2007.

FIGURE 3.5 Caltepec Fault zone, Sierra Mixteca, west side of the valleys. The velvety appearance of the mountains is due to forests of the columnar cactus *Neobuxbaumia tetetzo*.

stretching the Zapoteco terrane lying between them. The resulting depression—the Zapoteco terrane—began to subside, parts of it more than others, producing the Tehuacán graben and the valleys, which have collapsed relative to the Oaxaca and Caltepec Faults as the edge of the faults spread apart. This extensional stretching is responsible for the existence of the two valleys. The collapsing graben did not produce a depression of uniform width, for the Zapoteco terrane is far wider at its southern end than at the northern end, where it tapers to a point as it abuts the TMVB. The Caltepec Fault is better characterized as a zone rather than a visible structure, as is the case with the Oaxaca Fault.

Tectonic forces gave rise to the shape of the valleys and the surrounding mountains but tell us nothing about the composition of the rocks that compose the region. Mountain building, faulting at terrane boundaries, and collapse and degradation produced the valleys' climates and basic structure, but the different composition of the earth also figures prominently in the valleys' productivity. This includes the chemical composition and cohesion of the rock, which, in turn, determine the nature of the surface, weathering, soil texture, and, finally, the formation of the landscape as we see it. The varying composition of the earth also weathered and

FIGURE 3.6 Caltepec Fault zone, near Caltepec, Puebla.

decomposed into soils that in parts of the valleys are fragile and easily eroded—notably in the Mixteca Alta, as we discuss in chapter 7.

Much of the valleys are girded on each side by thick marine limestones. The portion of the larger Sierra Zongolica lying between the canyon of the Río Santo Domingo and the first major drainage to the north, known as Río Comulco, is often referred to as the Sierra Mazateca. The Río Comulco enters the Valley of Tehuacán just north of Coxcatlán. The range is named after the Mazatec people, who are its principal inhabitants. It is composed primarily of limestone and contains some of the longest and deepest limestone caverns in the world, shot through with underground streams, pools, and waterfalls. To the southeast, across the declivity excavated by the Río Santo Domingo, the Cueva Chevé (with the cave known as J2), in the vast limestones of the Sierra Juárez, may be more than 2,500 meters deep, perhaps the deepest in the world.[18]

Across the Cuicatlán Valley to the west, along the Río Xiquila and nearby Santa María Ixcatlán, are massive limestone formations, notably a series of karsts, which are hollowed-out limestone formations. The region is arid to semiarid, especially

18. Hose 1995, 108.

FIGURE 3.7 Box canyon upstream from Apoala Valley, where Mixtec legendry places the origin of the Mixtec people.

compared with the wet eastern slopes of the Sierra Juárez and Sierra Mazateca across the valley (recall that the weather masses from the Gulf of Mexico are mostly devoid of readily available moisture by the time they reach the Mixteca Alta), so the limestone formations erode at a different pace in these drier climates. Apoala, in a more mesic site south of Ixcatlán, is situated in a hanging garden surrounded on three sides by immense Cretaceous limestone cliffs. Upstream one kilometer from the town is a limestone cavern known to residents since pre-Columbian times as the Serpent Cave. Pre-Columbian codices make specific reference to the cave as where Mixtec ancestors emerged into the world.[19]

The Tehuacán-Cuicatlán Valleys have a deep history going back further than Pangaea. To the untrained eye, the valleys are a mosaic of steep mountain boundaries, rock outcrops, and loosely consolidated fill material from many different geological ages. A complex organization, however, remains in the rocks' structure. In a few, highly localized places in the southeast portions of the valleys, spectacularly ancient rocks are emplaced, some—mainly metamorphic—dating back one billion years, well into the Precambrian era. Other ancient rocks in the region of Los Reyes Metzontla and Atolotitlán are of Permian origin and contain a rich and ancient flora. The fossils include ferns, Gigantopterids—an extinct group proposed as a possible ancestor to all flowering plants—and broad-leaved gymnosperms related to the famous maidenhair tree, *Ginkgo biloba*.[20]

The upper reaches of Sierra Zongolica on the east show massive accumulation of mudstones, limestones, and sandstones of Late Jurassic to Early Cretaceous age. These sedimentary rocks include interbedded pillow lavas that formed at oceanic rift zones when moving lava encountered deep ocean water. They now lie nearly 2,000 meter above sea level. Southeast of Tehuacán City near Teotitlán de Flores Magón, extensive sandstones, limestones, and other calcareous rocks from the Cretaceous are dominant on both flanks of the valleys. In the lower pediments, Tertiary marine and lacustrine (lake) sediments are common. In all cases, rich assemblages of marine faunas dot the valleys, from littoral, shallow deposits with coastal seashells to deposits harboring fossils of large terrestrial and aquatic reptiles. The best example of these is around the town of San Juan Raya in the Zapotitlán drainage, where abundant and widely exposed Cretaceous fossils draw visitors from afar, and community members have joined in efforts to prevent illegal exploitation of the resource.

From the Middle Tertiary onward, lacustrine conditions replaced the former marine environments as the entire landscape was uplifted from the sea. These deposits occur at the bottom of the valleys and attest the formation of the large lake that

19. R. Williams 2009, 128.

20. "Pérmico de México," Virutas Paleobotánicas, December 14, 2022, http://www.geologia.unam.mx/igl/deptos/paleo/weber/florperm.htm.

FIGURE 3.8 Fossils, San Juan Raya, Zapotitlán Valley.

persisted until the recent formation of Santo Domingo canyon, which drained the valleys. The valleys continue to dry out, hydrostatically draining fossil water in a natural process. The exhaustion of the once-bountiful waters, however, has been greatly accelerated by groundwater pumping.

Most of the rocks in lower elevations are covered by Quaternary alluvial debris derived from alluvial fans and riverine deposition, as rivers transported their sediments into the basin and dropped them there. These formed the fertile soils where the valleys' first agriculturalists cultivated fields, which over the last couple of millennia have been terraced and contoured, tilled, and irrigated to provide the regional breadbasket.

A relatively recent geological development is the formation over the last sixty thousand years, possibly somewhat more, of large quantities of travertine, a form of limestone precipitated from hydrothermally heated springs.[21] Abundant deposits of travertine occur in veins around Texcala, on the slopes of the Valley of Zapotitlán and in hills that border the southwestern reaches of the Valley of Tehuacán. The Tehuacán travertine deposits produce workable stone that appears in an array of colors. The town of Texcala exists today thanks to the travertine, which local artisans call *ónix*. Over one hundred Texcalan families quarry or carve and polish

21. Michalzik et al. 2001, 88.

the native stone into sculptures, which they market all over Mexico and the rest of the world.

Some 30 kilometers southeast, large deposits of travertine on several hillsides are commercially exploited; the quarried stone is loaded onto trucks and exported to rock-carving workshops, primarily in Puebla and Mexico City, but internationally as well. A large deposit is being quarried immediately adjacent to Cerro Petlanco, a nondescript hill composed of travertine that has both botanical and archaeological interest (see chapter 8).

The hydrothermal activity that produced the travertine originates in a fault of Cretaceous carbonate settlements. Superheated water dissolves carbonates and associated minerals, which then precipitate out when the water reaches the surface, cools, and evaporates. The faulting, and perhaps the geothermal activity as well, appear to be connected to volcanism in TMVB. These deposits are probably recent additions to the Valley of Tehuacán.

The sixty thousand years during which the travertine has become emplaced are hardly the blink of an eye in geological time. In John McPhee's metaphor of deep time, if we represented the geological record for the valleys as the distance between our nose and the tip of our middle finger, human occupation of these ancient valleys is less than the distance erased by "a single stroke with a medium-grained (nail) file." Human occupation, however, has made a deep impact and altered the landscape for eons to come.

And so, climate and geology have combined to produce the broad variety of plant associations we find in the valleys. The variety of vegetation, the combinations of plants, and the abiotic environments seem endless. A subtle change in soil type, a slight increase or decrease in rainfall, a distinct location in relation to seasonal precipitation, a barely detectable shift in prevailing humidity, a few degrees alteration in angle of exposure to the sun, a few meters elevation change, a gradual or abrupt variation in the parent rock—any or several of these will produce a difference in which plants will grow, what combinations of plants will emerge, and, ultimately, what response will evolve to attract or repel a human presence. And these myriad nuances of geography give rise to a landscape of profound variability.

CHAPTER 4

The Pre-Columbian Valleys

We return to Cholula, Puebla, situated on the eastern side of the towering Sierra Nevada toward the southern rim of the Basin of Mexico. Although Cholula is located within the valley of Puebla, it sits near the edge of the Trans-Mexican Volcanic Belt, not far from where the ramp begins to descend into the Valley of Tehuacán. A millennium ago, it was a large city with an enormous pyramid—according to some, the largest pyramid by volume in the world—that dates back at least fifteen hundred years. Cholula also produced some of the finest ceramics in all of Mesoamerica at the time. The city's early ethnicity is elusive.

Cholula may have developed its own culture, but early in its history it shared elements with Teotihuacan, the huge city north of Mexico City that dominated central and southern Mexico for the first six or seven centuries of the Common Era. Somewhere between 720 and 750 CE, well after the collapse of Teotihuacan, alien military forces with mercantile interests conquered Cholula. From that hostile takeover, Cholula emerged as the dominant power on the great Mexican Plateau, replacing the dying economic power of Teotihuacan.[1] After that change of leadership, the great pyramid of Cholula achieved its largest dimensions. In the 1990s archaeologists discovered the tomb of a Maya priest or merchant buried within the Cholula pyramid, which gives us an idea of the great antiquity—and scope—of Cholula's international relations.[2] Indeed, Maya nobles may have been residents there in a Maya barrio.[3]

While we can only speculate about the nobles' arrival at the city from Yucatán, Chiapas, Guatemala, or Belize, the land route by which they arrived in Cholula probably involved climbing the gentle ramp along the easy road from Tehuacán, including

1. Pohl 1999, 14.

2. Suárez Cruz 1985; McCafferty 2001, 202.

3. McCafferty 2001, 206.

perhaps a sea route and river navigation along the Río Papaloapan. Whenever that priest or merchant, later carefully interred, or other nobles reached Cholula from south and east in Mesoamerica, they and their escorts must have been many. They traveled along a highway, road, or trail already ancient, as old as any traversed by Roman legions.[4] Their route led through Tehuacán, already a crossroads of merchants. The Conquest historian Toribio de Benavente, also known as Motolinía, wrote an extensive history and description of Mexico at the time of the Conquest while residing in a convent at Coxcatlán in the Valley of Tehuacán.[5] Motolinía mentions that Tehuacán formed "the frontier of many townships at the foothills of some mountain ranges, and from there many towns and provinces could be visited."[6] Then, as now, Tehuacán was a crossroads.

Aztec Soldiers, Merchants, and the Valleys

While the valleys have served as a trade and military corridor for many centuries, most of those campaigns have remained shrouded in antiquity. The conquest expeditions of Aztecs in the century before the Spanish Conquest, however, are documented in oral histories, in narratives recorded shortly after the Spanish Conquest of Mexico in 1521, and in pre-Columbian Aztec codices—scrolls depicting historical narratives.

During the century prior to the European Conquest, the Aztecs, also known as Mexica, had become the dominant power in the valleys. Beginning in the middle of the fifteenth century, they launched military offensives through Tehuacán, southward over Quiotepec Pass into La Cañada, and following various routes, spread into the central valleys of Oaxaca and beyond. One of these forays was the 1496 campaign of King Ahuízotl of Tenochtitlán, the Aztecs' capital city. His most celebrated military effort was an offensive against the Zapotecs of the Isthmus of Tehuantepec. The enemy kingdom lay in well-defended lowlands east of the central valleys of Oaxaca. The isthmus is Mexico's "waist," some 700 kilometers (435 mi.) southeast of the Aztec metropolis, across a series of mountains, valleys, and canyons. Despite the great distances the army needed to cross, Ahuízotl deemed the expedition so important that he supposedly organized and commanded it in person.

The threat posed by Ahuízotl's forces in Oaxaca hardly came as a surprise to Oaxacans. Armies from Tenochtitlán had previously engaged in sorties against Oaxacans and against Mixtecs northwest of the central valley. They had already engaged the Zapotecs of Tehuantepec in 1468–69 in an expedition led by Ahuízotl's

4. Long Towell and Attolini Lecón 2010.

5. Motolinía is a Náhuatl name meaning poor.

6. De Benavente 2014, 118.

older brother Axayácatl.[7] The result of that earlier engagement was the establishment in Tehuantepec of a supposedly peaceful Aztec garrison. It appears that the original terms of tribute imposed on the Zapotecs gave rise to their resistance to the Mexica's demands and resulted in Ahuízotl's punitive expedition.[8] During the intervening years Aztec couriers and spies had plied the roads between the isthmus and Tenochtitlán, while a few towns supported enclaves of Náhuatl (Aztec) speakers. Mexica *pochteca* (royal merchants) had expanded trade as they penetrated the southeast. These traders were mostly from the independent, market-centered city of Tlatelolco, immediately adjacent to Tenochtitlán.[9] Though their interests were primarily mercantile, they also passed along to the Aztecs military information that assisted the Aztec conquest of cities and states as the empire expanded. In addition to the pervasive merchants, previous Aztec emperors (and perhaps previous Nahua states) had ordered colonies of Nahua settlers to be established in conquered territories. Traces of these outliers of Aztec imperialism remain today.[10]

The Mexica nobility of Tenochtitlán and their aristocratic allies craved commodities unavailable to them in the lofty, cool Basin of Mexico. To obtain the tropical fruits native to the warm and well-watered valleys, they depended on locations south of Tenochtitlán, like Cuautla, Yautepec, and Oaxtepec, all conquered by the Aztecs under the rule of Moctezuma Ilhuicamina. As the empire expanded, they came to rely on the earlier-ripening tropical crops of Cuicatlán. The valley was renowned then as now as an excellent source of avocados (*Persea americana*), chicozapotes (*Manilkara zapota*), ciruelas (*Spondias purpurea*), mameys (*Pouteria sapota*), zapotes negros (*Diospyros digyna*), and the sweet but quickly perishable pitayas.[11] The latter

7. Hassig 1988, 176.

8. Marcus and Flannery 2000, 400. According to this tradition, Zapotecs made a secret deal with Aztecs, allowing Mixtecs to do the fighting while agreeing beforehand to allow Aztecs to establish a garrison.

9. Tenochtitlán and Tlatelolco were separate political entities living adjacent to each other on an island in Lake Texcoco, alternating allies and enemies, with a nearly symbiotic relationship. Tlatelolco was the home of most of the *pochteca*, who usually traveled under the protection of and with the blessing of the king of Tenochtitlán. He wisely recognized the importance of the merchants to the expanding economy of the Aztec capital, especially the luxury items craved by the Aztec nobility. See Davies 1973, 39.

10. In numerous towns in the Valley of Tehuacán Náhuatl is still widely spoken, probably a remnant of a Nahua diaspora as early as the eighth century CE. Elsewhere, the town of Maruata, on the coast of Michoacán, is a Náhuatl outlier as well. The town of Pochutla, near the southern (Pacific) coast of Oaxaca, was formerly home to Náhuatl speakers, although authorities disagree on whether it was an Aztec colony or simply a community of *pochteca*, Náhuatl-speaking merchants who survived the Spanish Conquest. Pipiles, descendants of Náhuatl speakers, remain in several communities in what are now El Salvador, Honduras, and Nicaragua. They may have preceded their distant relatives the Aztecs in the region by five hundred years.

11. Exchange between Cuicatlán and Oaxaca and Cuicatlán and Tehuacán is documented from the Perdido phase in Cuicatlán (roughly 400 BCE) by, among other things, the discovery in Oaxaca's central valleys of *Diospyros nigra* (zapote negro) remnants and other fruit fragments from La Cañada. Spencer and Redmond 1997, 602.

are fruits of the tall columnar cactus (*Stenocereus pruinosus*) that flourishes in the valleys but is absent in the Basin of Mexico. Even more, Aztec elite treasured exquisite salt harvested in the salinas of the Valley of Zapotitlán, a tributary of the Valley of Tehuacán. For the nobility, the valleys provided cochineal, the brilliant red dye; gold and silver jewelry from the mines and artisans of the Mixteca Alta; precious stones; parrot and quetzal feathers from southern forests; and from the tropical lowlands, cacao, whose value was such that the individual seeds constituted currency. The Zapotecs and Mixtecs of Oaxaca were also renowned for their fine cotton mantles, which Ahuízotl demanded in huge numbers.

But the Mexica's lust for spoils could not be satisfied with mere commodities. At the top of their list was a supply of captives from conquered peoples to be executed or recruited as slaves. The demand for prisoners seemed to expand with each new Aztec king. Small wonder that many tribute-paying captive peoples were eager at the time of the conquest to transfer their tribute payments from Aztecs to Castilians. From the earliest encounters, natives of the Americas realized that Spaniards could be cruel and ruthless, but they also noted that the Europeans did not appear to demand captives to be transported to the great city and did not levy increasingly heavy taxation. At least initially. Only later would the scope of European-inflicted atrocities become clear. In some localities the lust for gold and power eradicated entire native populations. Even during the conquest, Cortés ordered the Cholula Massacre, as we note below, and his lieutenant Pedro de Alvarado directed the Temple Massacre, where hundreds of men, women, and children were slaughtered. The list of European-inflicted genocidal activities is lengthy. Some argue that it is still expanding.[12]

Crucial to the Aztecs' campaign against Tehuantepec was the route the army would follow. The connection between the Central Plateau and southern Mexico, where the Oaxacans were located, lies in the linked valleys of Tehuacán, Puebla, and Cuicatlán, Oaxaca. This bridge between Mexico's two great landmasses had established the valleys' prehistoric significance for pre-Columbian civilizations for at least two millennia. In addition to the valleys' central role as a transportation corridor,

12. Aztec execution of prisoners of war has come under intense revision in recent scholarship. Spanish and later sources use the term "sacrifice"—a loaded word used to denigrate Aztec culture and imply primitive and barbarian practices. Archaeological evidence does not support the vast number of prisoners "sacrificed" as claimed almost universally by Spanish and, later, most international sources. Early Spanish adventurers in the Americas, their apologists, and their intellectual descendants vastly inflated the numbers and the nature of the Aztec executions to justify the destruction of Aztec and other American cultures while suggesting that Europeans were possessed of a superior morality and intellect. That many prisoners of war were executed there can be no doubt. The number pales to insignificance, however, when compared to the millions slaughtered, killed in pogroms of genocide, tortured, burned at the stake, brutalized, and later bombed by European and North American powers. Aztecs were far less violent than their European counterparts. See Restall 2018, 77–114.

they were repositories of virtually unlimited supplies of food for an invading army: cactus flowers and fruits, fresh in May through October, dried in other months. The columnar cactus called *tetecho* (*Neobuxbaumia tetetzo*) produces seeds that are easy to preserve, tasty, highly nutritious, and a good source of oils and protein. *Tetechos* grow in forests of untold millions of plants throughout the two valleys and, during their time of maturity, produce enough fruits to feed a small army. Other tropical fruits, cultivated for many centuries and mentioned above, abound there as well, and the tropical climate of La Cañada permits double cropping of corn. Irrigated fields may well have provided corn, beans, and squash in quantities sufficient to feed a small army, at least for a while.

Several military campaigns in the region were in full progress at the time of the Spanish Conquest of Mexico. Some expeditions continued after the arrival of Spaniards, while others were in progress during the 1521 battle between Mexica and Spaniards for control of Tenochtitlán, where Spaniards were busy spreading diseases and massacring resisters. Details are lacking, but news from Tenochtitlán of the defeat of the emperor must have resulted in chaos in the armed forces stationed abroad. The fall of Tenochtitlán shook up all of Mexico and Mesoamerica.

FIGURE 4.1 Zalea, the fruits of the columnar cactus *Neobuxbaumia tetetzo*. The seeds are rich in oil. The dried fruits are ground to make salsa. The seeds are ground to make *pipián*.

Before the Aztecs: A Brief Chronology of Civilizations

The Aztecs were late arrivals in the Mesoamerican military and economic drama. They followed the tracks of myriad peoples through the valleys. Below we provide in roughly chronological order a selected list of pre-Aztecan peoples of Mesoamerica who crisscrossed the valleys and who had attained sufficient cultural sophistication to construct monumental architecture and produce high-quality tools and artifacts. Their rise in architectural achievement was a clear sign of increased complexity of social organization and was accompanied by a corresponding rise in trade throughout the region. And, for our purposes, their commerce would have been conducted across the Cuicatlán and Tehuacán Valleys, along with less-traveled routes. Trade products from each have left behind traces of their presence in the great valleys. The reader should keep in mind that the periods discussed below cover three thousand years.

First, however, we note that archaeologists studying Mesoamerican prehistory distinguish among several periods based on the relative complexity of the societies of the period and their technological and cultural accomplishments. Each location had its distinct time scheme of development (often with time referred to in *phases* rather than *periods*). For example, in Oaxaca, archaeologists have divided the period from about 500 BCE to around 700 CE into three phases: Monte Albán I, II, and III. In the Valley of Tehuacán, the MacNeish team distinguished among what they referred to (from earliest to latest, 11,000 BCE to 1520 CE) as the Ajureado, El Riego, Coxcatlán, Abejas, Ajalpan, Palo Blanco, and Venta Salada phases.[13] In the Cuicatlán Valley Spencer and Redmond and other researchers divide the archaeological chronology into the Perdido, Lomas, Trujano, and Iglesia Vieja phases.[14] The following rough scheme for Mesoamerica in general reflects a consensus of scholars and compares general with specific locations:

Archaic: Prior to 2000 BCE (4,000 years before present). Societies were often nomadic, based on hunting and gathering. Early agriculturalists domesticated maize over several millennia and developed a diet based on maize, cultivated beans, squash, chiles, and a variety of vegetables, gradually developing into sedentary agriculture-based communities. In the Veracruz lowlands Olmecs incorporated manioc into their diet as well. In the Valley of Tehuacán, families grouped into hamlets and produced crude pottery.

Formative (Preclassic): 2000 BCE–200 CE (4000–1800 BP). Several societies formed chiefdoms. Class systems and division of labor appeared, along with

13. MacNeish 1964, 531.

14. Spencer and Redmond 1997.

administrative and religious centers and monumental architecture and sophisticated artwork. Maya, Olmecs, and Zapotecs developed writing, with accounts commemorating important events and persons carved in rock inscriptions. Sculptors of Teopantecuanitlán created elaborate friezes, and their engineers oversaw construction of stone-lined irrigation canals. Early towns and small cities emerged as trade and war stimulated international commerce and mutual defense structures. At the northern margin of the Basin of Mexico, Teotihuacan blossomed into a huge mercantile empire.

Classic: 200 CE–900 CE. Cities with social strata and division of labor as well as refined art and technology appeared. Monte Albán, and the Maya lowlands, especially Tikal and Calakmul, blossomed into full flower. Teotihuacan reached its highest influence. Trade became widespread, connecting the Basin of Mexico with Central America. Advanced astronomy and mathematics appeared. Monuments designed by Mesoamerican architects displayed their most elaborate and sophisticated achievements. By 700 CE all three empires had declined, and by 900 CE, none retained its status as a powerful city. In about 400 CE a volcanic eruption buried the city and small state of Cuicuilco, situated at the southern end of the Basin of Mexico. Cuicuilco was perhaps already in decline, but the outpouring of lava eliminated a competitor for dominance of Mesoamerica.[15]

Postclassic: 900 CE–1520 CE. The great urban centers had vanished and others, such as Xochicalco, had flowered briefly and were in decline. Tula, north of Teotihuacan, was on the rise, but it too flowered for only a couple of centuries. Only Cholula appears to have persevered through the Classic into the Postclassic. In Yucatán, the Puuc culture and the Itzás ruled much of the peninsula, though hardly on the international scope of Tikal. In the early fourteenth century, Tenochtitlán rose from the swamps of the southern Basin of Mexico and quickly overwhelmed virtually all competitors and underwent an explosive period of urban design and construction.

Early Organized Societies

Olmec: About 3,500 years ago, two and a half millennia before Ahuízotl's punitive expedition against Tehuantepec, Olmec civilization, which archaeologists generally consider the earliest in Mesoamerica, developed at the site of San Lorenzo in the lowlands of Veracruz and Tabasco.[16] The Olmec ascendancy, though politically limited to chiefdoms, flourished from around 1200 BCE until 500 BCE or possibly a couple of centuries later. Olmecs invented a written language and were both master builders and accomplished rock carvers, who exerted influence throughout

15. Pastrana and Fournier 2001, 290–92.

16. Recent discoveries at the site of Ceibal, Guatemala, suggest that early Maya cities may have preceded the earliest Olmec cities except for San Lorenzo. Takeshi Inomata, personal communication, 2014.

Mesoamerica. According to some, Olmecs invented the *comal* and the corn tortilla, which permitted great mobility to Olmec soldiers and traders.[17] With their increased mobility, technological sophistication, social organization, and, perhaps, military sophistication, Olmecs extended their influence well north of the Valley of Mexico and south into Central America. Some Olmec jade and pottery appeared in the Valley of Tehuacán as early as 1000 BCE.

Teopantecuanitlán: Nearly contemporaneous with the Olmecs of San Lorenzo, or perhaps a century later (1400 BCE–500 BCE), was the far smaller polity of Teopantecuanitlán ("the sacred place of jaguars"), located a good 400 kilometers west of the Olmecs in the Pacific drainage in southeastern Guerrero. Present knowledge is limited to a single site covering somewhat less than 2 square kilometers. Teopantecuanitlán exhibited sophisticated architecture, rock inscriptions, and irrigation via incised rock channels, all of which show strong Olmec influence or at least show that Olmecs and Teopantecuanes were actively exchanging ideas and goods at around 1000 BCE—undoubtedly across the Valleys of Tehuacán and Cuicatlán. Even though the site had a small population, probably little more than one thousand, and was probably ruled by a chieftain rather than an organized administration, Teopantecuanitlán is rich in items of high value imported from throughout Mesoamerica.[18]

Cholula: This impressive city constituted a large urban and political center at the southern end of the great Central Plateau of Mexico. Though its early ethnicity is unclear, Cholula's inhabitants probably spoke Náhuatl by the time of Aztec ascendancy and perhaps well before. It had developed complex social organization and monumental architecture during the mid-Classic (perhaps somewhat earlier), more than a millennium before the Aztecs of Tenochtitlán brought Cholula into their sphere of influence. Control over Cholula and the surrounding region probably changed hands several times over the centuries. The ethnicity of the Nahuas' predecessors is unknown. Cholula was overrun by peoples known as Olmeca-Xicalanca, probably originally from the coastal lowlands of Veracruz, around 750 CE (or somewhat later).[19] These newcomers conquered and drove out the Náhuatl speakers, who migrated southeast into Central America. Cholula was "liberated" from Olmeca-Xicalanca control by Toltecs of Tula, also Náhuatl speakers, at the beginning of the thirteenth century.[20]

Cholula's earliest signs of organized society began before 1000 BCE, nearly two and one-half millennia before the founding of Tenochtitlán. Construction on the great pyramid began about 500 BCE, during the early Formative, and continued for

17. Hassig 1992, 28; see also Cook 1947, 49.

18. Martínez Donjuán 2001, 200–201; Tucker and Montero 2008.

19. This group should not be confused with the Olmecs (Olmecas) from two millennia earlier.

20. Suárez Cruz 1985; McCafferty 2001, 202–6.

about 1,500 years with various stages and expansions. At the height of the Classic period, around the time of Olmeca-Xicalanca ascendancy in the city, the pyramid was the largest and most important structure in Mesoamerica, perhaps in all the Americas. The city may have housed barrios (neighborhoods) of Chochos, Ixcatecs, Mazatecs, and Popolocas.[21] As Teotihuacan declined, beginning in the fourth century CE, Cholula expanded rapidly. Archaeological data fix the final phase of Cholula development at the arrival of Toltecs around 1200 CE. By the time of the Spanish Conquest, only Tenochtitlán rivaled Cholula in wealth and influence.[22] At least part of its rise was due to its ability to control trade to the southeast, undoubtedly through the Valleys of Cuicatlán and Tehuacán, given its location not far from the northern end of the Valley of Tehuacán.[23] Cholula also became a revered religious center, and "pilgrims" appear to have made journeys from all over Mesoamerica to the great religious center, activity that would have reflected in the valleys as well.[24] Hernán Cortés went to great and brutal lengths to defeat—and massacre—the nobility of Cholula and destroy its ability to rule.[25] This resounding destruction of the rulers of a prominent city perhaps helps explain why the towns in Tehuacán and Cuicatlán Valleys surrendered to the Spaniards without a fight: Tehuacán, Teotitlán, Tecomavaca, Quiotepec, Cuicatlán, and nearby Tamazulapan, as well as Teposcolula in the Mixteca Alta.[26]

Tikal: In the Petén region of Guatemala, Maya selected the site of Tikal as ideal for trade, and by 800 BCE it had developed the trappings of urban civilization. Sometime earlier, the Maya kings of Ceibal, well to the south, were directing the construction of monuments and practicing sophisticated agriculture, writing, engineering, mathematics, and astronomy.[27] Tikal learned well from Ceibal. By Classic times, Tikal had become the most powerful lowland center of Mesoamerica, with an urban population as high as ninety thousand and a grandiose architectural landscape unmatched in the world.[28] Although its prominence rose and fell, it remained an important center until its demise around 900 CE. Tikal had trade links throughout Mesoamerica, via both oceanic and terrestrial transport.[29] Ceramics

21. Xicalanca are peoples of elusive origin thought by some to have been allied with Olmeca through mercantile interests and founders of the city of Cacaxtla, west of Puebla, in late Classic times (750–850 CE). See Hassig 1992, 101.

22. McCafferty 2001, 202–6.

23. Fiedel 1987, 285.

24. McCafferty 2001, 205.

25. Thomas 1993, 261–64.

26. Castillo Tejero 2004, 179.

27. Takeshi Inomata, personal communication, 2013.

28. Coe 1999.

29. Golitko and Feinman 2015.

from Teotihuacan, 1,300 kilometers distant, are abundant in contemporaneous Tikal, which indicates sophisticated means of transport for these items, which were fragile indeed.[30]

Teotihuacan: The Maya kings of Tikal and other lowland Maya cities may have carried on extensive trade with Teotihuacan, nearly 1,200 kilometers (746 mi.) northwest, by as early as 300 BCE. By the beginning of the Common Era, Teotihuacan had grown to become the largest city—and it truly was a city—in the Americas. It achieved a population of roughly two hundred thousand by 300 CE, larger than any European city of its time.[31] Its two main streets were 5 kilometers (3 mi.) long and met at a central plaza. Along the Street of the Dead, its principal avenue, were more than one hundred temples. The city included more than two thousand multifamily apartments and five hundred crafts workshops.[32] Its development involved complex trade networks throughout Mexico proper, Mesoamerica, and probably far to the north as well. It remained a city of artisans and traders among nations for more than half a millennium and was the dominant economic force in Mesoamerica. Its imperial status was short lived by pre-Columbian Mesoamerican standards but still far longer than any post-Roman European or post-Conquest American power has endured. Teotihuacan's success also extracted a heavy toll from its environs, as the production of colossal amounts of lime and mortar for construction required destroying forests for fuel and lumber and planting vast acreages of corn and other foods to feed the armies of construction workers.[33] Aztecs wondered at the enormous ruined city that lay some 50 kilometers north of Tenochtitlán. Its antiquity was greater for them than that of Mesa Verde in New Mexico for present-day North Americans.

As grand as the great city became, how and by whom it was ruled remains undetermined. No clear rulers have emerged, and a ruling class has not been clearly identified. Its merchants were international travelers, and its trade tentacles were extensive and imperialistic, but kings, emperors, directors, or even prime ministers have not emerged from the archaeological record. The huge numbers of residential buildings and their seemingly egalitarian style of construction suggest that Teotihuacan did not conform to the traditional archaeologists' view of Mesoamerican monumental sites. Nor did it seem to have maintained a military establishment in any traditional sense. What is clear, however, is that it was enormously influential in matters of trade and other commerce.

30. Fiedel 1987, 282–85.

31. Archaeologist Linda Manzanilla (2001) estimates the population as between forty thousand and two hundred thousand during its peak. Other estimates are even higher (Millon 1976, 212).

32. Marcus 2000, 59, 67, 60.

33. Cook 1947, 48; Ezcurra 1992; Barba and Córdova Frunz 1999.

Trade connections between Oaxaca and Teotihuacan are well documented, especially involving mica, which Teotihuacan imported from Oaxaca in great quantities. Other high-value items changed hands as well, apparently arriving via human caravans during early Classic times. Since the principal source of mica in Oaxaca was tightly controlled by the elite of Monte Albán, we must assume that broad trade agreements connected the two centers of power and that the valleys were the principal route of transport.[34] Pottery from Teotihuacan also appeared at Monte Albán.

The ethnicity of the founders and rulers of Teotihuacan is unclear. In the Late Classic Teotihuacan was overrun by Nahuas arriving from the northwest. From Teotihuacan, they went on to found Tula. Ultimately, they conquered most of the great cities and states of the Basin of Mexico. The advent of the Nahuas in the Basin of Mexico also forced Otomí to migrate north and west, where they remain today (and in so doing drove a wedge among Otomanguean speakers). On reaching the Valley of Tehuacán, the Nahuas forced Popolocas to the west and Mazatecs to the east (see map 2, in chapter 5).

Monte Albán: At San José Mogote, now a suburb of Oaxaca City in gently rolling country to the north, civic buildings appeared around 1000 BCE, monumental architecture by 500 BCE. Though never a major city and not yet a state, Mogote was a center of artisan production, including polished magnetite mirrors, which were traded throughout Mesoamerica. It was at Mogote or nearby that Zapotec writing developed. Researchers now believe it to be the oldest stone writing in all of Mesoamerica.[35]

Monte Albán, 15 kilometers south, gradually rose to power beginning around 500 BCE. It wrested control of the valleys from San José Mogote, forming in the process the oldest state in Mesoamerica.[36] The mountaintop site remained the locus of ruling power—and enormous ceremonial and administrative structures—in Oaxaca for over one thousand years, until after around 700 CE, perhaps as late as 850 CE. It became the dominant economic power of all southern Mexico (probably excluding the Maya lowlands, but perhaps there as well), with a sophisticated system of writing, monumental architecture, elaborate artwork, and a highly stratified society, including an army. Monte Albán appears to have conquered La Cañada (specifically Quiotepec and Llano Perdido) between 300 and 200 BCE, and the Zapotecs controlled the Cuicatlán Valley for well over four hundred years. While Monte Albán constituted an imperial center, dictating foreign policy from the heights of the citadel, its population was barely a tenth that of Teotihuacan. Even including its metropolitan area, it was home to a population that appears never to have exceeded thirty

34. Manzanilla et al. 2017.

35. Spencer and Redmond 2004, 176; Spencer 2003, 11187.

36. Spencer 2003, 11187.

FIGURE 4.2 Restored raised platform at San José Mogote, where a civilization developed long before Monte Albán.

thousand. Marcus Winter suggests that Teotihuacan may have exerted military control over Monte Albán during the Classic period.[37]

Xochicalco: This site sits atop a large hill in western Morelos, north of Monte Albán but south of the TMVB. The hilltop, which commands a fine view of the surrounding countryside, was home to a large and sophisticated culture. Xochicalco represented one of several new economic and military powers that emerged as Teotihuacan faded. It was a relative latecomer in the succession of Mesoamerican powers but became an influential trading center, with elaborate and massive structures, all protected by carefully wrought defenses. The military aspect of the site perhaps was a response to the power vacuum following the demise of Teotihuacan. The citadel was defended via a series of walls and moats, reflecting that during its period of influence, armed attacks were expected, and military experts and engineers were prominently engaged in complex schemes to ward off invaders. Xochicalco also employed a highly trained cadre of engineers and scientists, whose building skills and knowledge of astronomy were state of the art at the time. Engraved glyphs demonstrate that the inhabitants had mastered writing. Xochicalco's ascendancy lasted for only about two centuries, between 700 and about 900 CE, when it was

37. Winter 2007, 34.

FIGURE 4.3 Monte Albán, with Building J in the foreground. In the background on the upper left is Cerro Atzompa, a monumental site built somewhat later than Monte Albán but clearly in visual contact with the former. From Atzompa the site of Suchilquitongo, a similar mountaintop side some 15 kilometers north, is also visible.

attacked, burned down, and abandoned.[38] While Oaxacan influence in both style and artifacts—primarily ceramics—appear at Xochicalco, it is difficult to establish the connecting routes that their cultural interchanges might have followed.[39]

Tula: North of Xochicalco about 150 kilometers and north of Tenochtitlán in the state of Hidalgo, Tula (also called Tollan) emerged as a regional power at roughly the same time as Xochicalco, in the early eighth century CE.[40] Its inhabitants came to be known as Toltecs and, it is widely believed, were Nahua speakers recently arrived from the northwest. Tula was home to an obsidian manufacture industry, as Teotihuacan had been five hundred years earlier, and it became renowned as the major source in Mesoamerica for the manufacture of obsidian tools. It was also a mass producer of ceramics and at the same time a major importer of pottery, much of it from Central America. It became the largest city of central Mexico, home to sixty thousand to eighty thousand residents. At its height, its influence extended as far away as Yucatán.

38. Hassig 1992, 104–7.

39. Negra 1989, 96.

40. Mastache and Cobean 2001, 269–74.

That influence—military and cultural—continued throughout Mesoamerica until about 1200 CE, when Tula fell and was sacked by Chichimecs, northern folk with less imperial and material pretensions, the Goths of Mexico. Despite Tula's sudden decline three centuries before the arrival of Spaniards, its history is consistently documented in Aztec codices and in post-Conquest oral histories. Aztecs viewed Tula's ruins with reverence and shipped pillaged stone sculptures from the ruined site to Tenochtitlán for public display. They appear to have considered a pilgrimage to Tula a spiritually enriching journey. Tula may have influenced the development of Chichén Itzá in the Yucatán Peninsula, or vice versa. Some authorities believe emigrants from Tula, probably Náhuatl speakers, overwhelmed Cholula and assumed political control of the great city after Tula itself fell to Chichimecs. According to this theory, they remained in power there, resisting Aztec assimilation until the Conquest and Cholula's ignominious defeat by the forces of Hernán Cortés.

Tenochtitlán: Aztecs, who claimed to be the successors of Tula, founded Tenochtitlán in 1324–25 CE. One hundred years later they were the uncontested imperial rulers over all central Mexico west of Michoacán. Their armies conquered states and peoples east and south, well into Mesoamerica. They wielded power over the region for little more than a century, but the effects of their empires were permanent. At its height, Tenochtitlán had a population of a quarter of a million people, making it one of the world's largest cities, certainly the most splendid and, by most accounts, the cleanest. Eyewitness descriptions of the city from Europeans portray it as the architectural wonder of the world in the early sixteenth century. Its vast markets hummed with activity, displaying goods from throughout Mexico and Mesoamerica and even beyond.[41] The king himself officially sponsored its merchants, the *pochteca*. *Pochteca* traveled thousands of kilometers to obtain the high-quality goods Aztec nobility craved. In turn, the Aztec kings provided the merchants with military support whenever necessary and encouraged subjected peoples to accept the goods produced by Aztecs and their allies in trade for luxury items. The splendor of Tenochtitlán can be directly linked to the success of the Mexica kings and their soldiers in exacting fearsome terms of tribute from conquered nations.

Place names throughout Mesoamerica (except in most of Michoacán and in Maya country) reflect Tenochtitlán's overwhelming military and political control of a vast region, yet the Aztecs as a people did not remain in many of the locations beyond the Valley of Tehuacán that bear Náhuatl names. Why did Náhuatl place names persevere in Cuicatlán-Oaxaca-Tehuantepec-Chiapas, but not the Náhuatl language?

41. Early descriptions of the wonders of Tenochtitlán are abundant. Especially useful accounts can be found in Davies (1973). Perhaps the best description is to be found in chapter 8 of Bernal Díaz de Castillo, *The True Discovery and Conquest of Mexico*. Various translations are available. Díaz was an eyewitness to some, but by no means all, of the Spanish Conquest of Mexico.

FIGURE 4.4 Bell tower of the church at Altepexi, Valley of Tehuacán, a Náhuatl town. The represented soldier is strongly reminiscent of military statues from the Toltec site of Tula.

It may be that there simply was not ample time for the Aztecs to establish social domination over such a large area. The seventy-five years or so of Aztec warfare may not have provided historical exposure and social pressure necessary to establish a *spreading front* of Náhuatl. In other words, Náhuatl never became a language so dominant and backed by a political power so influential that the spoken language extended permanently beyond areas where Nahuas lived. Conceivably as well, the forces of linguistic expansion may have been in place and abruptly terminated: troops and even colonists dispatched to conquer and settle may have been recalled to Tenochtitlán by Moctezuma II at the time of Conquest or just prior, thus removing the linguistic pressure on the spreading front.[42]

The Mixteca: Mixtec civilization represents an unusual Mesoamerican phenomenon: although Mixtec architects, engineers, stonemasons, jewelers, and potters were as sophisticated as any in greater Mesoamerica, no single Mixtec site dominated the region. As an example, we highlight here Yucuita, a partially excavated site that once occupied two square kilometers (roughly 700 acres) in the cool highlands of the Mixteca Alta of northwestern Oaxaca, just beyond the limits of the La Cañada drainage. We could also have chosen different sophisticated urbanized sites nearby, such as Huamelulpan, Cerro Jazmín, or Monte Negro, but none of these gained more than temporary regional ascendancy. Nearly two millennia before Tenochtitlán, earlier than the rise of Monte Albán, Mixtec chiefdoms began to coalesce into small polities. Indeed, Spores and Balkansky argue that the Mixteca Alta, not the Olmec region, may represent the "mother" of Mesoamerican cultures.[43] Yet none of these chiefdoms exerted international muscle. None developed a multistratified society. Still, as a group they were influential throughout Mesoamerica well beyond their highland locations. By the time of Monte Albán's rise, Yucuita, perhaps the most prominent of the polities, had already been occupied for a millennium, predating Monte Albán by several centuries.

Yucuita developed at the southern base of a symmetrical hill visible for many miles around, nearly straddling the boundary between La Cañada drainage and the Pacific drainage. We were surprised to find the site unmarked except for some local signs. Some minor restoration has taken place, but the surface is archaeologically depauperate, only scattered but intriguing remnants of its former importance jutting from the surface. Yucuita remained a civic-ceremonial center for hundreds of years, though its civic functions were probably secondary to its ceremonial importance. Yucuitans developed sophisticated ceramics and architecture, but no military notoriety. The official or officials of Yucuita ordered the construction of

42. For articulation of the linguistic terminology *spreading front* and *residual front,* see Nichols 1992, 22ff.

43. Spores and Balkansky 2013, 52.

FIGURE 4.5 Yucuita, a major site in the Mixteca Alta near the western edge of the Valley of Cuicatlán. It flourished prior to Monte Albán, as early as 1000 BCE. Yucuita and most Mixtec archaeological sites lack monumental architecture.

monuments, elegant and finely crafted but only on a comparatively modest scale, and presided over a quasi-suburban agrarian countryside without a notable urbanized zone.

This lack of dominant cities per se is characteristic of the Mixteca Alta.[44] As Monte Albán's power increased, however, Yucuita became more tightly urbanized, and defensive walls appeared, which suggests that peasants were gathering in the urbanized area for protection, a local response to Zapotec militarism and demands for tribute.[45] So, while Mixtecs of Yucuita and other sites matched Zapotecs for artistic and cultural achievements, they did not contest Zapotec military might or imperial impositions until the decline of Monte Albán after about 600–700 CE. Even though we have no evidence of military conquests from Monte Albán, and though

44. The absence of large urbanized centers and monumental architecture in the Mixteca Alta resulted in relative obscurity and lack of attention from archaeologists until the latter half of the twentieth century. The region's population and ceremonial centers were—and continue to be—diffusely located. Today virtually the entirety of the highland region constitutes a massive archaeological site.

45. Marcus and Flannery 1996, 377; Balkansky 1998, 467; Kowalewski et al. 2009, 336.

Yucuita remained independent for centuries, Monte Albán held sway, and Yucuita remained under Zapotec imperial influence for over a thousand years. Indeed, as Zapotec military and imperial power expanded, Mixtecs responded by flocking to the towns, which became increasingly defense oriented but never highly urbanized.

Mixtec prehistory is documented in codices, pictorial depictions of history, religion, and genealogy, a distinct and most useful source of chronology. Though most were destroyed by Catholic clergy, enough remain to provide genealogical history. Using codices in combination with archaeological findings, Mexican archaeologist Alfonso Caso traced the genealogy of Yanhuitlán, a *cacicazgo.* This term referred to a group ruled by a cacique—a Taino word appropriated by Spaniards throughout the Americas to designate kings and chiefs and an occasional queen, called a *cacica.* The town, home to an immense sixteenth-century Dominican convent, is seated near the western edge of La Cañada and dates from as far back as 620 CE.[46]

Considering the region's cultural longevity, along with its extraordinary creativity in ceramics production, it seems odd that Mixtec artifacts are relatively absent in the archaeological record of La Cañada, the land of the Cuicatecs and their Zapotec conquerors, while abundant in polities to the north. This may be explained by the routes connecting the Mixteca Alta and Monte Albán to the south and Cholula and the Basin of Mexico to the north. As archaeologist Blas Castellón points out, the most logical route between the Mixteca Alta and the salinas (salt works) of Zapotitlán in the Tehuacán Valley drainage lay to the west of La Cañada, passing through what is now Zapotitlán Salinas, and arriving at Tehuacán to the northeast, thus bypassing the valley route.[47] The Mixtecs apparently saw little need to descend into La Cañada to communicate to the north or to the south. Their trade route demonstrates the multiplicity of routes that crisscrossed the entire region. We understand the Mixtec's reluctance to venture to the valley, having several times traversed the steep descent into the often fierce heat of La Cañada.

Zapotecs in La Cañada

Monte Albán's control over Oaxaca expanded around 500 BCE, extending northward to the lowland valleys, though no farther. Caso in 1947 suggested that Zapotecs conquered and controlled the La Cañada region between late Monte Albán I

46. Caso 1963, 313–18; Spores 1967, 59. Lest one question the reliability or accuracy of such long traditions based mostly on oral renditions, Rustom Bharucha reports that eminent Indian folklorist Kolmal Kothari documented castes whose sole responsibility was retaining genealogical records of their patron. These sources could easily recount genealogies up to twenty generations. Others could relate as many as fifty. Bharucha 2003, 28–34, and personal communication 2008.

47. Castellón 2006, 62.

(600–200 BCE) and late Monte Albán II (200 BCE–200 CE).[48] Subsequent archaeological studies have confirmed his findings. In 1976 Joyce Marcus proposed that an inscription depicting a conquest on Building J at Monte Albán referred to the Zapotec overthrow of Cuicatlán.[49] Since Building J was constructed during Monte Albán II, the conquest must have occurred prior to that time. Excavations by Spencer and Redmond in the Cuicatlán region since the mid-1970s have established the antiquity of Zapotec occupation at around 300 BCE. Zapotecs sacked the Perdido phase site of Llano Perdido near Santiago Dominguillo, killed many of its occupants, forced relocation of the survivors to a nearby hillside site, and mounted a host of human skulls on a rack for public viewing as a reminder to the hapless inhabitants. The formerly rich cultural and social life of the Cuicatec victims of Zapotec violence was reduced to a colonial system of agricultural peonage for the benefit of the elite of distant Monte Albán.[50]

The bloody conquest demonstrated that Monte Albán's military muscle was rising, and lesser folk, especially those who might exhibit hints of resistance, were vulnerable to attacks from the Valley of Oaxaca. Monte Albán expanded in power and in population, and by the beginning of the Common Era, the mountaintop site had a population of about seventeen thousand.[51] A much larger metropolitan area, rather densely populated by commoners or peasants who lived in the valley bottoms, surrounded the huge mountain city. In addition, to the northwest lay the already densely populated Mixteca Alta. An indeterminate but sizable population of Cuicatecs and other groups inhabited the Sierra Juárez and the coastal lowlands to the northeast. Tututepec, a Mixtec settlement near the Pacific coast, was a modest regional center during the Late Formative, but once Zapotec power waned during the Classic period, Tututepec expanded to become the dominant power of the coast, the lower valleys, and the foothills. During Postclassic times, perhaps four hundred years after the decline of Monte Albán, Tututepec reigned as an important commercial center.[52]

These Oaxacan urban centers were of modest size in comparison with Teotihuacan, which by 450 CE may have had between 150,000 and 250,000 inhabitants.[53] (A thousand years later, Tenochtitlán had a population of well over 200,000.)[54] Still, Monte Albán's highly developed administrative and military structures, though on a

48. Jiménez Moreno 1966, 12–13; other authors use different terminology and alternate dates and periods. See also Spencer and Redmond 2001.

49. Spencer and Redmond 1997, 13.

50. Spencer and Redmond 1997, 12.

51. Marcus and Flannery 1996, 139.

52. Levine 2011, 22–39.

53. Coe, Snow, and Benson 1986.

54. M. E. Smith 2005, 411.

FIGURE 4.6 Close view of Building J, Monte Albán, completed prior to 200 CE. According to archaeologists, one of its many carved panels depicts the Zapotec conquest of Cuicatlán.

smaller scale, permitted the rulers to exert imperial control over lesser principalities and towns. Their reach extended primarily to the north and east.

The Zapotecan capital never exerted sufficient strength to be considered more than a regional imperial military power, yet the rulers were able to control traffic between the Tehuacán and Cuicatlán Valleys and probably over the Mixteca Alta as well. Even earlier than Monte Albán, somewhere around 1000 BCE, pre–Monte Albán Zapotecs had traded well beyond their sphere of influence in Oaxaca, exchanging mirrors produced in San José Mogote just north of the city of Oaxaca with consumers in the valleys of Morelos, some 500 kilometers northwest.[55] As the scope of the empire that ruled from the summit of Monte Albán expanded (and the influence of San José Mogote proportionally waned), the rulers indulged their craving for the tropical fruits of Cuicatlán, trading their ceramics for produce from the valleys.

55. Marcus and Flannery 2000, 365. Exchange items excavated in the site at San José Mogote include *Spondylus* shells, stingray spines, conch shell trumpets, and turtle shell drums, all of which were imported from ocean sources. Some of the *Spondylus* (thorny oyster) may have come from Ecuador (Macías Goytia 2000, 378; Shimada 1999, 350–517, esp. "Mesoamerican-Northwest South American Connections," 430–36).

They also saw both a benefit and a threat in the valleys functioning as a highway connecting central and southern Mexico and recognized the need to control traffic through La Cañada. And so, according to the archaeological findings of Spencer and Redmond, they invaded and conquered La Cañada. Early in their history they traded goods with Olmecs from the coast of Tabasco and Veracruz, and later, with Teotihuacan, north of Mexico City, and to a lesser extent with Maya from Yucatán and Guatemala. Early in their ascendancy, Zapotecs probably traded with Teopantecuanitlán, Guerrero, the small ancient center influential even then.[56] In addition, a dozen small principalities, or *cacicazgos*, minor states whose rulers held sway over a limited area, participated in the Zapotecan trade network. Over the centuries, the route through the valleys saw many thousands of merchants, soldiers, and peasants representing an array of clans, families, statelets, rulers, and governments, passing back and forth. At various stages, emissaries of Teotihuacan, Cholula, El Tajín (a Totonacan site in western Veracruz), Xochicalco, El Zapote (an Olmec site in eastern Veracruz), Yucatán, Chiapas, the Maya highlands and lowlands, and numerous other nations in varying states of power relationships followed the fast route from southeast to northwest and back again.

As this brief survey of Mexican and Mesoamerican population centers demonstrates, as far back as 1000 BCE the reach of urban buying and trading power, sophisticated production of commodities, and large urban populations meant that trade possibilities were widespread between and through the valleys and back and forth between southern and central Mexico. By the time of the rise of Monte Albán, Zapotecs had become key players in Mesoamerican commerce. Kings and princes saw the advantage of regulating and controlling trade to their benefit and sometimes to the benefit of their subjects as well.

At the time of the Aztec campaigns against Tehuantepec, Zapotecs themselves had long since ceased to constitute the principal military power in the central valleys of Oaxaca. By 300 CE they had withdrawn from La Cañada. After reaching the zenith of their power in Monte Albán around 600 CE, the thousand years of Zapotecs' supremacy (never as widely pervasive as that of Teotihuacan) waned. Over the next couple of centuries, the Oaxacans found themselves buffeted by increasingly aggressive Mixtec *cacicazgos*. Nearby sites in the central valleys, such as Jalieza, Tilcajete, and Dainzú, attained urban status and were probably competitive with Monte Albán in size and economic (perhaps even military) status.[57] By 750 CE Monte Albán was

56. The routes connecting the Gulf lowland Olmec sites such as La Venta or San Lorenzo with Teopantecuanitlán, some 500 kilometers distant, would have passed through La Cañada and the Valley of Tehuacán. At that time Monte Albán was a growing power. Description of the archaeology of Teopantecuanitlán is largely the work of the Mexican archaeologist Guadalupe Martínez Donjuán.

57. Balkansky 1998, 474.

in or on the verge of military and economic collapse.[58] Though the site continued to be inhabited, its imperial influence virtually vanished.

The Rise of the Mixtecs

Meanwhile in the Mixteca Alta, the power of the *cacicazgos* was expanding. Population density had been high for more than a millennium: during the Formative period, it appears to have exceeded that of the Basin of Mexico and of the Valley of Oaxaca.[59] The Mixtecs very early achieved social and artistic sophistication, producing ceramic creations of breathtaking beauty and complexity. The *cacicazgos* were often at war with each other, however, and their small empires, if we can call them that, remained only peripherally influential in regional politics until toward the end of the Classic period and the decline of Monte Albán.[60] After that time, Zapotecs could no longer control their neighbors militarily. Mixtecs, apparently somewhat unified by that time, gradually filled in the vacuum. Though they never conquered the Zapotecs, they mustered sufficient armies to force the Zapotecs to relinquish and share power in the central valleys. Such important centers in the Mixteca as Apoala, Monte Negro, Tilantongo, Tlaxiaco, Tututepec (located in the Pacific coastal lowlands), Yanhuitlán, and Yucuita had long rivaled Monte Albán's artistic and architectural accomplishments, even if they lacked its military capability. By the eighth century CE they deemed themselves capable of challenging the fading Zapotec power. As proof of their emerging power, they seem to have participated in the overthrow of the rulers of Cholula as well. Still, dominant power among the Mixtec *cacicazgos* shifted frequently, and at the time of the rise of Tenochtitlán's imperial power in the late fourteenth century, most of Mesoamerica, especially south of the valleys, consisted of *micro-states*, small regional or local principalities or statelets in a constantly changing mosaic of political power, none of it centralized to any degree.[61]

Archaeological studies in portions of the Mixteca Alta point to a lengthy hiatus in the development of monumental architecture and artistic creativity in the population centers, a lull roughly between 600 CE and 1300 CE.[62] During this lengthy and perplexing lull, the large-scale projects characteristic of the Classic period seem nearly to have ceased, along with any trends toward urbanization. It is as if the Postclassic period skipped the Mixteca almost entirely. At the end of this time, however,

58. Marcus and Flannery 2000, 391.

59. C. A. Smith 2002, 121.

60. Marcus and Flannery 2000, 372–73.

61. Hunt 1972, 242.

62. Kowalewski et al. 2009, 328–29; Spores and Balkansky 2013, 90.

the region experienced an explosion of creativity in Mixtec arts—ceramics, jewelry, and codices. Nor did the apparent freeze on urban concentration prevent *cacicazgos*' military forces from moving into Cholula in the eighth century or from spreading their influence into the central valleys of Oaxaca. The caciques may have shifted their budgets into military spending rather than funding capital projects and artisans.

By the time of the Aztec aggression in the mid-fifteenth century, however, the *cacicazgos* had regained their momentum toward building urban areas, as the prominence of Coixtlahuaca and Tlaxiaco demonstrate. At the time of the Spanish Conquest, Mixtecs, whether as one unified force or several, appear to have controlled most, if not all, of Oaxaca's central valleys, in some cases exercising joint control with Zapotec rulers through intermarriage of nobility or other alliances, in others, ruling over the Zapotecs.[63] The Aztecs must have confronted—and dealt with—the Mixtec statelets as they worked their way through the central valleys en route to Tehuantepec, where they faced remnants of the Zapotec empire.

As the influence of Mixtecs expanded toward the end of the first millennium CE and during the perplexing hiatus, many Zapotecs appear to have filtered eastward over the centuries, seeking refuge in the hot lowlands of the isthmus. Their descendants flourish there today, speaking a Zapotecan language largely unintelligible to Zapotecs of the lofty Valley of Oaxaca to the west and featuring a matriarchal society. Some Zapotecs remained behind in the central valleys, their military power diminished, forced (or choosing) to form alliances with Mixtecs. Others relocated to different semiurban locations such as Zaachila, some 30 kilometers south of Monte Albán, and other small polities. No Mixtec center appears to have developed the military power to assert itself above the others or to imitate the former reach of Monte Albán. Zapotecs seem to have retained a semblance of power in Mitla, near the eastern limits of Oaxaca's central valleys, or even shared it with Mixtecs, and in Zaachila, though with a vastly diminished power status. At no point do Mixtecs appear to have exerted firm control over the central valley of Oaxaca. Zapotecs, despite a general diaspora, continued to constitute the majority of the inhabitants there. Mixtec power appears to be mostly absent from La Cañada as well, possibly because at that time Cuicatecs were sufficiently organized so as not to be easily dislodged. The Mixtecs also seem to have found communications and trade north via the Valley of Zapotitlán to be more productive and less prone to conflict than via La Cañada. Zapotecs were still the reigning power in the isthmus at the time of Aztec military campaigns.

Late in the Postclassic, the Mixteca Alta population exploded, along with its burst of artistic creativity and agricultural bounty. Some of their towns were robust market

63. Paddock 1966, 375.

centers. Thus, we can account for the Aztecs' eagerness to conquer the major urban centers and milk them for tribute, slaves, and captives to execute, and the Spaniards' subsequent haste to convert and exploit the most populous region in Mesoamerica.[64] The Dominican priests' success in overseeing construction of three enormous convents in the Mixteca Alta, structures that dwarf any others in Oaxaca, testify to the region's relative affluence.

Roads and Trails

> The Camino Real itself . . . followed the old Aztec trade route to Oaxaca via the Cañada, which Cortez, Alvarado, and other Conquistadores used in their movements to Veracruz and the Etla region of Oaxaca.
>
> —EVA HUNT, "IRRIGATION AND THE SOCIO-POLITICAL ORGANIZATION OF CUICATEC CACICAZGOS"

Long before the rise of Monte Albán at around 500 BCE, economic powers with trading interests had made use of the valleys. Without convenient access through this alley, movement of people and goods was arduous. Traveling overland involved slogging through hundreds of kilometers of the marshy and jungle-overrun, mosquito-infested Gulf coast lowlands of Veracruz and the abrupt foothills of the Sierra Juárez on the north and east, or the interminable steep inland mountain ranges and nearly impenetrable thorny Pacific coastal forests of Oaxaca and Guerrero on the south and west. Trade by boat was ongoing along the Gulf of Mexico and the Pacific coast, with connections to inland markets. The Maya excelled in maritime and riverine trade, but the cultures of the upper valleys mostly traded by land.[65] Commerce by boat had the advantage of far greater load capacity than human porters but was subject to the hazards of tide, wave, and flood, plus the obstructing mountains beyond.

Olmecs maintained vigorous trading ties with distant markets as early as 900 BCE and surely maintained roads from their homelands near the Gulf of Mexico in southern Veracruz and eastern Tabasco.[66] For the next half millennium they established a far-flung network of trade.[67] For peoples from the lofty plateaus and valleys of Mexico, Puebla, and Oaxaca, the climate of the lowlands would have been to them what it is now, startlingly rich and biodiverse but unpleasantly hot and humid, lush

64. Spores and Balkansky 2013, 91–93.

65. McKillop 2017.

66. A good, if somewhat dated, summary of the Olmecs is Coe and Diehl (1980).

67. Olmec influence extended in the Formative (Preclassic) period from San Lorenzo Tenochtitlán and La Venta in Tabasco as far north as Tlatilco in the Mexico City area to Chalchuapa in El Salvador.

with jungle vegetation and tropical pests, infections, and parasites. While communications between the Olmec homelands and Teopantecuanitlán in northeastern Guerrero were established as much as three thousand years ago, those routes, predominantly east–west, must have been more difficult than the northwest–southeast route between Mexico City and Oaxaca via the Tehuacán and Cuicatlán Valleys.[68] Zapotecs, who knew the value of the route through Cuicatlán and used it as their zone of influence, still used the La Cañada route until 900 CE, long after Monte Albán's demise as a center of power.[69] These ancient pathways continued to be travel routes throughout the Spanish Conquest and the viceroyalty period. They remain a vital element in contemporary travel, as the twentieth-century highway system between Mexico City and Oaxaca followed the old routes. The modern toll highway, constructed in the early twenty-first century with heavy machinery and graced with imposing bridges, carved out a new, faster route.

Where were these trails and roads? A trail or pathway is an informal and traditional means of walking or riding between two points. A road or roadway is a deliberately established means of traveling between two points. The distinction between the two is not clear-cut, but paths are usually narrow and, in hilly terrain, winding only when necessary, with a much steeper grade resulting in a shorter distance between places. Paths usually allow no more than two people to transit side by side, while roads are usually wider. Straighter pathways, although present in pre-Hispanic America, became common with the arrival of horses and carriages. The predominance of pre-Columbian paths over roads in Mesoamerica was simply the consequence of the absence of beasts of burden and, in Cuicatlán and Tehuacán Valleys, the absence of a centralized government.

In pre-Columbian Mesoamerica, human porters carried out all transport. In Aztec times, *tamemes*, human carriers who traveled in groups, transported all goods. The regularity of their passage should suggest the existence of roads as opposed to paths, but we have not been able to identify traces of any such formal route. In any case, a path with heavy use may evolve into a road, and a little-used road may degenerate into a mere pathway. Paths or trails may be improved by the placement of rocks, berms, crossings, or even bridges, while roads may fail to conquer obstacles and thus contain "unfinished" sections.

Spanish speakers do not always linguistically distinguish between paths and roads. The word *camino* can be used for either.[70] A formally designated highway in New

68. Martínez Donjuán 2001, 200–201.

69. Morrill 2008, 892. Morrill's date of 900 CE is unexplained. The town of Cuicatlán (an Aztec name) may not have existed until after the conquest of La Cañada by Aztecs. See Roque 2012, 18.

70. The words *senda*, *sendero*, and *vereda* in Spanish can be used to refer to paths or trails. A well-used pathway connecting two settlements is usually referred to as a *camino*.

Spain was called a camino real, a royal highway. Whether or not the connecting route through Tehuacán and Cuicatlán Valleys constituted a camino real in post-Conquest Mexico, the route was certainly used heavily, as authorities in central Mexico maintained close ties with Spaniards in Oaxaca, and Oaxaca, known as Antequera to Spaniards, became an encomienda of Hernán Cortés.

As if to underscore the informal nature of the roads, sixteenth-century Franciscan friar and historian Motolinía describes pre-Hispanic paths as rough, narrow passages that overcame barriers by using ladders made of poles with cutout steps where half a foot could fit.[71] Ropes helped the climber or descender to pull and keep equilibrium. The description of using poles, ropes, and rock linings gives credence to the straightness of the paths. Mexican anthropologist Francisco del Paso y Troncoso asserts that paths were "as narrow and straight as measured by level and ruler . . . they did not bend [from the straight line] a point" ("tan angostos . . . tan derechos y medidos por regla o nivel que . . . no lo torcían un punto").[72] Pathways served as roadways.

We have not identified an improved roadway through the valleys from pre-Hispanic times, nor have we managed to pinpoint the probable route to the central valleys of Oaxaca via the Río de las Vueltas from Santiago Dominguillo to San Juan del Estado, near Oaxaca City. This is not entirely surprising, since in Middle America neither paths nor roads required heavy maintenance except in areas and seasons of heavy rainfall. Much of the easiest routes would follow the floodplains of the rivers, and trails or roads would be erased as the riverbed meandered. Still, an informal route could easily evolve into a more formal route merely because it was the most topographically logical one and became the favored route through constant use. A force of only a dozen or so men carrying burdens will string out in a long line, their collective footfalls leaving a noticeable trail or path.[73] Larger groups will leave behind even greater lasting traces, and their effects amplify their tracks well out of proportion to their numbers. An army of sixty thousand, supposedly the size of the Aztec expedition against Coixtlahuaca in 1458, would leave nearly permanent reminders along the route, creating for all practical purposes a road, while the northward tramping of thousands of captives, roped closely together so that they had to march in formation, would compact the soils for generations to come. Nothing the pre-Columbian peoples of the Americas possessed could match the incisive effect on soils of wagon wheels and beasts of burden, but ongoing usage, especially in the semiarid valleys,

---

71. De Benavente 2014, 137. "Otros pasos muy ásperos subíamos por escaleras . . . eran hechas de un palo sólo, hechas unas concavidades, cavado un poco en el palo, en que cabía mitad del pie y sogas en las manos." These devices are still commonly used in rural Mexico to climb rooftops, cliffs, or trees.

72. Del Paso y Troncoso 1905–, 16.

73. Hassig 1992, 21.

would have left the principal trade routes well marked and readily visible.[74] The best routes connecting trade points would have been easy to discern, if arduous. And restocking outposts, ersatz convenience stores, would have been located at regular intervals along the way. Routes would have been selected based on access to water and flat places on which to pitch camp, as modest as they might have been.

Similarly, if small groups regularly use a trail, the route will become easily recognizable unless it is erased by flood, landslide, or ongoing heavy precipitation. While the early users of land routes of Mesoamerica must have traveled in small numbers—the logistics of supplying larger groups with food would have been daunting—repeated usage of the routes would mark them indelibly.[75]

Still, the absence of detectable extended roadways seems odd. It contrasts with pre-Columbian roads in Andean South America, in the U.S. Southwest, and in Yucatán. Beginning in the thirteenth century CE, the Incas maintained thousands of miles of roads (and many bridges composed of woven fiber), usually at least 3 meters wide, surfaced with dressed stone, and stepped with precisely cut and measured rocks. Llama trains then and now used the roads to transport merchandise and food over long distances. The constructed roadways vastly improved communications through the vast reaches of the Inca empires and afforded the rulers rapid movement of messengers, armies, and supplies. The Incas invested enormous expense in these all-weather roads, many of which still function perfectly.

Well to the north, in what is now New Mexico, the elite of Chaco Canyon oversaw construction of many miles of roads, invariably straight, including elaborate staircases when cliffs and ravines intervened. These mostly connected with nearby Chacoan sites, but some extended north as well, perhaps more than one hundred miles. Those that connected Chacoan sites were well maintained and engineered, making communication among Chacoans more efficient.[76] They may have served a religious or symbolic function as well as supporting intercultural reinforcement and trade.[77]

The Maya of the Yucatán Peninsula also created hundreds of miles of elevated, generally arrow-straight paved roads called *sacbeob* (singular *sacbé*: white way) during the Preclassic and Classic periods.[78] These elevated structures were a necessity for military action and widespread commerce in Yucatán. The bedrock of the peninsula is rough limestone; the vegetation grows back rapidly during the rainy

74. Compared to horses, mules, and burros, llamas and alpacas, the domesticated camelids of the Andes, have soft hooves and are far less prone to compact soils, instead leaving incised trails in their wake.

75. Traffic throughout Mesoamerica would also have ebbed and flowed relative to the cycles of planting corn and incidences of famine. Until well into the twentieth century wars and battles in Mexico were often put on hold or delayed so that peasants could plant, cultivate, and harvest their corn.

76. Lister and Lister 1981.

77. Noble 2004, 83.

78. See, for example, Villa 1934, 162.

season; and much of the terrain is unvaryingly flat: no natural landmarks visible from afar exist over thousands of square miles. We can testify that getting lost in the Yucatec forest (especially on a cloudy day) can be a harrowing experience. The lords of the great Maya centers had a powerful vested interest in constructing and maintaining the *sacbeob*. Without the publicly administered highway department, maintenance of trade and political domination would have been quickly compromised.

The Incan and the Chacoan road systems were constructed over a couple of centuries' time by administrative order in areas politically and economically controlled by the road builders. In Tehuacán and Cuicatlán Valleys, no single group ever appears to have been able to exercise control over regional traffic that traversed both valleys. No polity existed that could have been charged with construction and maintenance of highways or could have afforded the expense of constructing permanent roads. The roadways that existed would have been spontaneous reactions to the landmasses that lay between the travelers and their goals. Until the Zapotecs constructed the fortress on the Cerro Quiotepec and asserted control over the passage between the north and the south, we must assume that the roads belonged to no one and were freely adopted and adapted by merchants and soldiers alike as they saw fit.

Paving was not as vital for the routes through the valleys connecting Monte Albán with Teotihuacan, Tenochtitlán with Tehuantepec, or Tabasco with Guerrero as it was in Yucatán. Pathways and roadways in the hill and mountain country must have been arduous to traverse but easy to follow once they were established and used regularly. Except for in the humid lowlands of the Gulf coast, the compaction of soils from the pounding of feet, the removal of obstructing stones from the roadway to one side or the other, and the constant traffic would maintain the routes free of obstructing vegetation. The surfaces were largely benign, compared with the irregular and sharp-edged limestone of the Yucatán. Furthermore, the countryside is hilly to mountainous, so that landmarks are readily visible and getting lost less probable.

Large armies of the day were hardly flying companies. While a single runner can travel as much as 60 kilometers a day, pre-Columbian soldiers of Mesoamerica were seldom capable of achieving more than roughly 20 kilometers, even with forced marches.[79] They had no beasts of burden or wagons and thus could not move quickly or over long distances. The cumbersome logistics of supplying weapons, food, and other supplies rendered surprise attacks on a large scale by such armies nearly impossible. Soldiers weighted down with their weapons—offensive and defensive—could hardly march speedily over long distances with the added burden of food and

79. For calculations of army march speeds, see Hassig 1992, 21–23.

bedding supplies. Throngs of porters were required just to support the fighting force. And the porters themselves needed to eat and sleep. The protocols of war of the time also required that military officers and nobility justify their status by defeating opponents in hand-to-hand combat. Their positions of privilege were announced to enemy and ally alike by their lavish accoutrements, and thus they were also weighted down with ponderous regalia.

The logistics of long-distance travel meant that local sources of food and supplies were essential along major routes. Around 200 BCE, more than a millennium and a half before Ahuízotl's expedition, the Zapotecs of Monte Albán occupied Cerro Quiotepec, located at the junction of the Valley of Tehuacán and La Cañada, and constructed a massive fortress overlooking the rivers' junction. That installation became the northernmost outpost of Monte Albán's supremacy, and the Zapotecs appear to have operated something like a customs and immigration station for passersby in all directions, thus controlling any northern threats to their privileged trade position. Arable land near the fortress is quite limited, however, and the numbers of fighters and their families stationed for duty would probably have readily exceeded available food supplies. Near San José Tilapa, some 30 kilometers north of Quiotepec, Purrón Dam had been a source of irrigation water for crops since the Late Formative, well before the construction of the fortress on the Cerro de Quiotepec.[80] The role of products of the Purrón irrigation district as exchange in trade along the Mexico–Oaxaca trade route needs to be examined, but part of the dam's rationale must have been the provisioning of travelers along the route and, it would seem, of the sentries of Quiotepec and their associates. Coxcatlán, north of Tilapa, is a site of great antiquity, and its relationship with the Presa Purrón was of great prehistoric importance.[81] Leaders from Coxcatlán must have had ongoing discussions, probably not altogether friendly, with the Zapotecs of Quiotepec concerning the limitations on trade imposed by the great fortress.

From a practical standpoint, then, merchants, traders, and military personnel needed food and shelter along the lengthy journeys between the Basin of Mexico and southern Mesoamerica, and a modest travel infrastructure must have existed to accommodate their needs.[82] If trade or military activity decreased, the reprovisioning sites along the route would have experienced lesser demand and provided

80. For a description of Purrón Dam, see chapter 6.

81. Coxcatlán is mentioned in the Códice Chimalpopoca ([1992] 2019, 10:49) as a site settled by Toltecs who abandoned Tula at the time of its downfall. Later it would be conquered by Aztecs. The Coxcatlán Cave is the source of much of Richard MacNeish's findings on the antiquity of human occupation and agriculture in the Valley of Tehuacán.

82. The construction of the Interstate Highway System in mid-twentieth-century United States proved catastrophic for many hundreds of towns and small cities as the new highways bypassed their business and visitor infrastructure. Almost overnight these businesses languished, and many thousands of them disappeared.

fewer services, resulting in greater hardships for travelers and lean times for local providers. Traffic along the valleys must have constituted an important factor in the local and regional economy. Trade has its multiplier effect.

The Aztecs' ability to transport large armies required extensive planning and a compliant population along the routes. Residents along those routes would have become accustomed to seeing passersby or modest military squadrons (though armies were probably a terrifying novelty) and quite probably would been involved in trade with them. If soldiers pillaged settlements along the route, the inhabitants would have moved elsewhere to a more secure location, hence abuse of native populations who were needed for supplying moving armies would have been self-defeating. Even so, while populations were sparse and provisions in short supply along lengthy stretches of the road, Aztec armies levied tribute on pueblos along the route to Tehuantepec and severely punished those who resisted. The imperialists imposed their demands on chiefs and leaders, however, rather than on individuals. This interference in free trade must have aroused resentment among the victims of the nonvoluntary taxation.

Aztecs probably expended considerable effort to keep the routes open and free from robbers or military impediments. Oaxaca was a source of tribute, trade, luxury commodities, and captives. It also was an irreplaceable conduit to Central America and markets beyond. The conquerors established forts at Coixtlahuaca and Sosola (located midway between Coixtlahuaca and Monte Albán) to keep trade routes well stocked and tribute flowing in the proper direction.[83] As we shall see, Mexica kings retaliated with vengeance when their stream of tribute was interrupted. Small wonder that once Spain had broken the back of the Aztec military might, resistance elsewhere in the Aztec empire was comparatively lame.

Over a thousand years before Ahuízotl's reign but long after the waning of Olmecs' influence, that is, roughly the first through seventh centuries CE, the power brokers of Teotihuacan rose to power. Teotihuacan extended the reach of their influence—their merchants—well into Central America and far to the north, reaching Mexico's desert northwest and the midwestern United States. On an international scale, Teotihuacan's influence dwarfed that of Monte Albán. Indeed, Teotihuacan lured a community of Zapotecs to settle in their city, some 600 kilometers northwest of Monte Albán: John Paddock has described a Zapotec neighborhood in Teotihuacan in which expatriate Zapotecs produced some of the artifacts for which they had become internationally renowned.[84] Somewhat later, following the peak of Teotihuacan, Maya followed Olmecs in establishing routes of exchange well

83. Reina Aoyama 1998.

84. Paddock 1966, 375.

into the north, perhaps even beyond Teotihuacan. They, too, dwelled in barrios in Teotihuacan and in turn hosted Teotihuacanos in barrios of Tikal.[85]

From the Basin of Mexico, which over a span of 1,500 years included the empires of Cuicuilco, Teotihuacan, Tula, and Tenochtitlán (among others), a route east to the valley of Puebla, perhaps through Cholula, then southeast through the Valleys of Tehuacán and Cuicatlán permitted large trains of merchants and far larger armies to move with relative ease.[86] While the corridor required passage over mountains more than 3,000 meters in elevation, the passes were not especially onerous. Water was readily available along the route, and temperatures, though chilly in the high passes east and south of Tenochtitlán and torrid in the lowest parts of the valleys—reaching as high as 49°C—were tolerable and escapable. From the earliest years, furthermore, food was probably available from local sources along the entire route.

The paths and roads of the valleys formed part of the arteries on which the prosperity of kingdoms and empires depended—and their armies. Travel between Mexico City and Oaxaca remained arduous and slow until the completion of a railroad between Tehuacán and Oaxaca by a British company in 1893. A previous attempt by a company headed by retired U.S. president Ulysses S. Grant to build a railroad connection had failed, but the British company completed construction in just over three years. The route followed the Valley of Tehuacán, crossed into La Cañada near Quiotepec, and then climbed from La Cañada via Tomellín Canyon and its drainages to emerge into the Valley of Etla near El Huitzo.[87] The train flourished with federal assistance until the mid-1990s, when the Mexican government eliminated all government funding to its railroads. The highly subsidized construction of the high-speed freeway made travel between Tehuacán and Oaxaca by bus several hours faster than by train. In 2010, torrential rains in the upper basin of Río Grande and Cañada Chica produced massive flooding downstream. Large sections of the railroad were washed out. Repairs would have amounted to many millions of pesos and would probably have required a change in the alignment in some area. Without public announcement, the trains stopped running. Residents of El Parián, an isolated railway stop in the upper Tomellín Canyon, relate that they, along with a score of travelers en route to Oaxaca, waited for days for the arrival of the train, which had appeared regularly for more than one hundred years. It never came. Most residents

85. Schele and Freidel 1990, 162.

86. Except, that is, for the last few kilometers, where the route departed the Río de las Vueltas and ascended into the Valley of Etla, north of Oaxaca City. The international influence of the autonomous polity of Cuicuilco, an early sophisticated competitor of Teotihuacan, located next to the Escuela Nacional de Antropología e Historia in Mexico City, was stifled by the eruption of the volcano Xitle in the foothills of Ajusco volcanic range in about 400 CE. See Adams 1991.

87. Hernández 2014.

FIGURE 4.7 Río Tomellín downstream from El Parián. Note the railroad bed. The railway was abandoned early in the twenty-first century.

FIGURE 4.8 El Parián, a railway village in the Río Tomellín drainage that has become a ghost town since the railroad was abandoned.

shrugged and departed for more promising locations. El Parián, once a bustling town, is now a virtual ghost town, with only a few families remaining in residence. We searched there in vain for a place to eat. We settled for tortillas with salt for spice. The era of the railroads in the valleys is over. The canyon of Tomellín is no longer a transportation route.

The Aztecs in Oaxaca Before Ahuízotl

Monte Albán declined as an imperial power after around 600 CE. Mixtec principalities, which had long since responded to Zapotec militancy by concentrating in defensible urban areas, emerged with greater influence, and Mixtecs appeared in increasing numbers in the central valleys. Much of what we know of the Mixtecs is gleaned from Mixtec codices—painted narratives—and transcribed accounts of Aztec warfare against Oaxacan principalities in the fifteenth century. At that time, nearly nine centuries after the decline of Monte Albán, power in the central valleys and highlands of Oaxaca was distributed among several Mixtec and Zapotec *cacicazgos*, none of them a dominant power, none of them capable of asserting itself over the others for more than a few years. In La Cañada and the mountains of the northeast, Cuicatecs and possibly Mazatecs had reasserted themselves in the power vacuum but were also divided into small to medium-sized *cacicazgos*.[88]

Aztecs appear to have been involved in separate military actions against Oaxacans (apparently both Mixtecs and Zapotecs) several decades prior to Ahuízotl's expedition against Tehuantepec. Moctezuma I had launched a punitive raid against rebellious Zapotecs and/or Mixtecs of the Oaxaca Valley in the mid-1450s, annihilated their city, massacred many, and took numerous prisoners as captives and slaves.[89] They arrived at Tenochtitlán just in time to be executed at the dedication of Moctezuma's newly constructed temple. The size and ethnic composition of the conquered city of Oaxaca is not clear from historical descriptions, but clearly it was far smaller than the seventeen thousand or so residents that occupied Monte Albán at its pinnacle of power a millennium earlier.

The Aztecs hoped to use their destruction of Oaxaca as an example to other principalities, but that strategy apparently failed.[90] The logistic problems posed by

88. Cline 1963, 270–97; Hunt 1972, 211. Residents of the towns of Atatlahuca and Zoquiápam in La Cañada identify their villages as Chinantecan, not Cuicatec, as Hunt believed. Chinantecs may have been a factor in determining regional dominance of one *cacicazgo* over the others.

89. Durán (1588) 1994, 228–31.

90. Moctezuma I, or more probably his brother Tlacaelel, declared that the only human flesh suitable for consumption by Aztec nobility would be that of enemy Náhuatl-speaking cities, including Tlaxcala and

maintaining distant garrisons were formidable. Fray Diego Durán does not mention Mixtecs as living in the Monte Albán sector of Oaxaca, although they frequented the valleys of Oaxaca by that time. It also appears that whatever vestiges of the Zapotec empire remained by the time of Aztec invasions, no other major cities had survived for the Aztecs to conquer and brag about. The valleys and nearby mountain civilizations of Oaxaca were balkanized, not in the sense that they were fragments remaining from the breakup of a larger state, but in that they were a collection of statelets or *cacicazgos*, none of which clearly dominated the others.[91] At the time the Aztecs arrived, the long-abandoned mountaintop city of Monte Albán lay in ruins high above whatever urban area persevered.

Not long after Moctezuma I defeated Oaxaca, Tlacaelel, his brother and the true power behind the throne of the early Aztec kings, announced a program to repopulate the vanquished city, which, it seems, had been destroyed. Under Tlacaelel's orders, six hundred married men and their families from Tenochtitlán and allied cities were dispatched to settle in Oaxaca, families from the different cities forming their own barrios amid the native Zapotecs and Mixtecs.[92] If the figure of six hundred families is correct, the population of the newly founded city would be at a minimum in the vicinity of 2,500 inhabitants (plus current native residents), still a far cry from the former city many times that size. At any rate, according to the historical tradition, in this way Oaxaca came to have a strong Aztec presence. Durán explains that the colonization was not simply for military purposes. Aztec legions destroyed cities and their crops, leaving no workforce to produce the tribute the king hoped to enjoy, hence the need for colonists to plant, labor, and produce, replacing the savaged populations.[93]

Durán describes the Aztec colonization of Oaxaca as follows:

> Motecuhzoma [I] called his cousin Átlatl whom he made governor of all those people and ordered him to arrange the city of Guaxaca in such a way that the Aztecs would form a barrio of their own, the Tezcocans another, the Tepanecs still another, and the Xochimilcas and all the other groups also their own quarters. The leaders and those men that the governor deemed the wisest were to be made dignitaries and given positions of authority. That city was to be controlled with the order and harmony char-

Huejotzingo, traditional enemies of Tenochtitlán. All other human flesh was barbarian and unworthy and inherently distasteful to the palates of Aztec nobility (Durán [1588] 1994, 233). One may thus understand the Tlaxcalans' willingness to join forces with Cortés to defeat Tenochtitlán. Whether or not this Aztec dietary restriction consoled the non-Nahuas has not yet been made clear.

91. Marcus and Flannery 2000, 394; see also Kowalewski et al. 2009, 340.

92. Durán (1588) 1994, 236.

93. Durán (1588) 1994, 344–45.

> acteristic of Tenochtitlán. . . . When they reached Guaxaca, these people re-founded and rebuilt the city. Following the instructions of the king each group was given its own barrio. The officials who had accompanied the governor to Guaxaca returned to Tenochtitlán and reported to the king that the city of Guaxaca had been newly built, was now at peace and in good order.[94]

Tradition has it that Tenochtitlán directed the rebuilding of the city in 1486, but it probably began prior to 1450. Several barrios of Oaxaca City still retain the Aztec titles: Chapultepec, Mexicapan, and Xochimilco. Barrio Jalatlaco is of Tlaxcalan, not Aztec, origin. The Tlaxcalans were Nahua-speaking mortal enemies of the Aztecs. The Spanish Crown used large numbers of Tlaxcalan families to colonize throughout New Spain as a means of providing friendly and loyal settlements in the periphery.

Spaniards appear to have brought Tlaxcalteca with them to Tehuacán, and the new arrivals enjoyed high rank and special privileges in their new home. Paredes Colín notes, "All of those with the last name *Santiago*, and there were many . . . were the chiefs of Tehuacán and it appears they were originally from Tlaxcala."[95] Tlaxcalans were probably sent to Oaxaca for much the same reasons: to colonize the city and to ward off general uprisings against Spanish rule.

Spaniards merely carried on the tradition of colonizing established by the Aztecs, sending out settler groups of Tlaxcalans. King Ahuízotl had perpetuated the colonizing system established by earlier rulers, directing several thousand married couples from Tenochtitlán and other allied cities to repopulate cities in Guerrero that had been conquered and razed by Aztec forces.[96] U.S. proslavery interests used the same strategy of encouraging large numbers of immigrants in the 1840s to settle in Texas to help gain its independence from Mexico and then its annexation as a U.S. slave state.

The linguistic evidence for the historical colonization of Oaxaca by Aztecs or previous Nahuas appears to have vanished almost completely. The community of Náhuatl speakers closest to Oaxaca is the hamlet of Cerro Verde, located along the Oaxaca/Puebla state line, in the vicinity of Teotitlán de Flores Magón in the Valley of Tehuacán. Place names in Náhuatl remain widespread, however. The names Teotitlán del Valle, a Zapotecan Indigenous community east of Oaxaca City; Mitla; and Tehuantepec are just three of many dozens. Cuicatlán is an Aztec name.

It must have been a nearly new population that constituted the rebuilt polis. By that time many, if not most, Zapotecs had long since abandoned the Monte Albán

94. Durán (1588) 1994, 237.
95. Paredes Colín (1910) 1953, 68.
96. Durán (1588) 1994, 344; Hassig 1988, 208.

metropolitan area, some of them migrating en masse to the Isthmus of Tehuantepec, others to towns in the sierras beyond the easy reach of the Mixtecs or Aztecs, still others occupying lesser sites in the valleys. Some traditions make the city of Zaachila, south of Oaxaca, the repository of Zapotecan royal descendants, while Aztecs took over the newly rebuilt city of Oaxaca.[97] Zaachila was also a Mixtec stronghold.[98] King Axayácatl, Ahuízotl's brother and royal predecessor, had already attacked and defeated Tehuantepec in 1468–69 and returned to Tenochtitlán with captives.[99] It seems odd that despite repopulation by Aztecs, no traces of the spoken Náhuatl language remain in the city of Oaxaca.

Some of the Mixtec *cacicazgos*, themselves representing *ñuu* (autonomous social states) perhaps thousands of years old, may have chosen to trade with the Aztecs and pay them tribute rather than fight, much as they did with Zapotecs during Monte Albán's ascendancy. By the late fifteenth century, various minor cities had developed a brisk trade with Tenochtitlán. Some of the Aztec merchants and traders had taken up residence in entrepôt towns, one of which was the city of Oaxaca, now possibly under nominal Mixtec control, Zapotecs having dispersed to Mitla, Zaachila, and the Isthmus. The old city's location in the central valley of Oaxaca, then as now, made it a crossroads of commodities from the highlands, lowlands, Gulf of Mexico, and Pacific Ocean. Other trade centers in Oaxaca included the *cacicazgos* of Coixtlahuaca (probably of Chocho ethnicity at this time), Yanhuitlán, and later, Tlaxiaco. It must be remembered, however, that none of these centers were major urban areas. Instead, they were important political and trade centers presiding over a rather dispersed agrarian population.[100] Tenochtitlán stationed *pochteca* in these busy centers as well as in the city of Oaxaca. In addition to their role in acquiring treasured commodities for the Aztecs, the traders introduced imports from the north, mostly from Tenochtitlán, but from other settlements as well. It appears that local merchants were required to accept the imports from the north, whether they wanted them or not, in exchange for the treasures of the Mixtecs. When the colonies resisted, as was the case with several Mixtec towns, the Mexica responded with severe repression.

The earliest well-documented campaign against peoples of the Mixteca occurred in 1458.[101] King Moctezuma I, the first imperialist Aztec king, organized an expedition to punish Coixtlahuaca. This ancient town lies in the cool uplands of the Mixteca Alta of northwestern Oaxaca, in the farthest southwestern watershed of the Valley of Tehuacán. It is situated on a plateau 1,000 meters or so above the Río Grande, the great

97. Marcus and Flannery 2000, 391–92.

98. Paddock 1966, 371.

99. Hassig 1988, 176.

100. Spores 1967, 101.

101. Códice Chimalpopoca (1992) 2019, 51:33.

river of the Cuicatecs that lies due east. The area is now impoverished, barren, and eroded, but five hundred years ago, it was prosperous and verdant. The expedition was apparently a different campaign from the one Moctezuma had sent to punish Tehuantepec in the same year. Coixtlahuaca, a city then governed by the fierce Chocho people (and perhaps by Mixtecs as well), was a renowned center of manufacture, especially of gold and silver jewelry, but also of decorated gourds and an exquisite fabric woven from spun rabbit fur.[102] Some traditions consider the rulers of Coixtlahuaca to have been Otomanguean descendants of the rulers of Tula, who migrated south when the city was defeated in the twelfth century.[103] Coixtlahuaca was home to considerable numbers of *pochteca* from Tenochtitlán and its allies. They traded with the Mixtecs or Chochos for the luxury items so craved by the nobility of Tenochtitlán to the north. That privileged class had seen its numbers and affluence constantly increased under the expanding power of Aztec kings, the accrual of tribute from colonies, and the subjugation and exploitation of the peasantry.[104] Extraction of surplus value from the labor of others was not an exclusively European or capitalist invention.

The Chochos must have resented the imperious Aztecs in their midst. Some incident caused the Coixtlahuaca caciques to crack down on the *pochteca* and scapegoat them for one reason or another, unleashing what must have been pent-up fury at *pochteca* arrogance. Apparently under orders from their king, Atonaltzin (a Náhuatl name), the Chochos executed 162 of them and drove the remaining merchants out of the city.

When news of the mistreatment of Moctezuma I's representatives and allies reached him, he sent an army that marched directly on Coixtlahuaca and was defeated. Furious, Moctezuma called on his allies and vassal states. He mustered a huge army—estimates place the size of the army at two hundred thousand soldiers and one hundred thousand porters.[105] After careful preparations, the multinational expeditionary army marched through Tehuacán, located at that time on La Mesa, and easily defeated the Popolocas of the town, forcing them to relocate from La Mesa to more or less the present location of Tehuacán and to accept the Aztec name for their kingdom.[106] With Tehuacán as a supply center, the Aztecs proceeded to march on Coixtlahuaca. Many of the soldiers trembled with fear due to the reputation of

102. Durán (1588) 1994, 182ff. Chochos, who speak an Otomanguean language (related to Mixteco and Zapoteco), are greatly reduced in number. As few as two hundred speakers remained in the year 2000. Most of these now live in Santa María Nativitas, a small town located a few kilometers southwest of Coixtlahuaca.

103. Códice Chimalpopoca (1992) 2019, 10:49, 51:37.

104. For a chronology of the rise of Aztec absolutism and the subjugation of the *macehaules* (commoners), see Padden 1967, 93–97.

105. Hassig 1988, 166. "So many soldiers were recruited . . . that they seemed to cover the earth"; Durán (1588) 1994, 183. See also Paddock 1966, 228–36.

106. Paredes Colín (1910) 1953, 43–44.

FIGURE 4.9 Coixtlahuaca, Mixteca Alta, an ancient Chocho town. Note the severe erosion and barren slopes contrasting with the cultivated bottomlands left of center.

the Chochos as "devilish savages." The legions passed southward through Valley of Tehuacán, arriving at a side valley that empties from the west just north of the gentle pass that leads into La Cañada. From there it was but a day's march uphill to the land of the Chochos. The Aztec armies could easily be provisioned from the rich farmlands and orchards of La Cañada and the irrigated milpas of Valley of Tehuacán.

The Chochos of Coixtlahuaca, despite their reputation for ferocity, were no match for the Aztec legions, who were disciplined and well trained. Coixtlahuaca offered only token resistance. The army from the north slaughtered the Chocho fighters. The leaders of Coixtlahuaca, seeing their cause was lost, prostrated themselves before the Aztec leaders and agreed to provide tribute to Tenochtitlán, including captives to be marched north as slaves or executed at Aztec ceremonies. The Aztecs killed the king, made a prisoner of the queen, whom they took to Tenochtitlán to be presented to Moctezuma. He found her too fat.[107] And thus a permanent Aztec military presence was established in what is now the state of Oaxaca, only to be supplanted by Spaniards some seventy years later. The Europeans simply took over the Aztec position, leaving the tribute system intact for the time being.

107. Códice Chimalpopoca (1992) 2019, 51:40.

If the Aztec legions ventured far south in the Valley of Tehuacán they must have wondered at the ruined temples of Los Cues, Tecomavaca, and Teotitlán and the magnificent and colossal fortress atop the Cerro de Quiotepec, much as Roman legions viewed with wonder the ruins of Etruscan and Greek temples and cities. Some of the soldiers might even have brought home small souvenirs from the remains of exquisite Zapotec ceramics that must have littered the site.

Even more disastrous for the Mixtec region than the defeat of Coixtlahuaca was the conquest of Tlaxiaco in 1502, more than forty years later. It lies southwest of Coixtlahuaca in the pine-forested highlands of the Mixteca Alta and remains a vigorous city of Mixtecs and several related Indigenous cultures. Before and after the Spanish Conquest, Tlaxiaco was a major commercial center. Like the Coixtlahuacans and Tehuantepecans, Tlaxiacans openly defied Aztec rule and flaunted their defiance and independence. Even worse, they intercepted and stole some of the tribute the now-docile Coixtlahuacans had collected to be sent to Tenochtitlán and killed some of the Chochos who were bearing tribute to the Aztecs on their backs. Moctezuma II, the most arrogant of the Aztec kings (he had proclaimed himself a deity, and no one seemed to argue the point), was furious about the theft of his booty and launched a genocidal attack on the *cacicazgo*. The Tlaxiacans put up no defense, and a huge Aztec army went on a rampage, killing half the inhabitants, including women, and marching a host of the survivors off to Tenochtitlán, another round of slaves and victims for Huitzilopochtli.[108] Moctezuma II had elevated the Hummingbird God to the position of supreme deity and correspondingly upped the god's demand for sacrificial victims, reputedly numbering in the tens, perhaps hundreds of thousands. The captives from Oaxaca passed through the Valley of Tehuacán (perhaps first through the Valley of Zapotitlán), bound and tied together, with wooden collars around their necks, on the grueling—and probably fatal—march and arrival at Tenochtitlán.

Without ready access and communication between Mexico's great central basin and the fertile, pleasant valleys of Oaxaca and Chiapas to the east, Aztecs could not have moved their massive armies over long distances to protect trade and tribute. Their wars were only the latest in a long history of struggles, some military, others economic, between the two power centers for control of resources and exploitation of distant markets. The strategic significance of the valleys remained the same over the centuries: Whoever controlled the access points in the valleys also controlled the prime trade and war route connecting southeast and northwest, the Basin of Mexico and Central America, which were home to the great population centers of Mesoamerica. And whoever controlled the route dictated the flow of commerce

108. Durán (1588) 1994, 470–73.

between the prosperous regions of the north and south and all their related networks. Naturally occurring obstacles—hills and passes—through the valleys were easily overcome, impeded only by minor intervening mountains. Political impediments were a different matter.

La Cañada and the Cuicatecs

As Mixtec *cacicazgos* rose in power west of La Cañada during the Postclassic period (i.e., after roughly 900 CE), several Cuicatec *cacicazgos* of varying size and power developed around irrigation works in La Cañada and on the mountainsides to the east.[109] Some may have originated or expanded into the western slopes as well. These survived Aztec domination and remained in place throughout the sixteenth century under Spanish rule. They may have competed for trade and power with adjacent Mixtec *cacicazgos*. Caso notes that in the early sixteenth century, Mixtec families settled as refugees (or perhaps invitees) in Cuilápam, a Zapotec suburb of Zaachila, just south of Oaxaca City. They had departed (perhaps fled) the town of Almoloyas in Tomellín Canyon on the western slopes of La Cañada. They first sought refuge in Yanhuitlán, a Mixtec center some 100 kilometers northwest of Monte Albán, believing Yanhuitlán would offer them protection from aggressive Cuicatec neighbors. From there they emigrated to Cuilápam, apparently not feeling secure even in Yanhuitlán.[110] Rincón Mautner believes the migration was more voluntary, a result of intermarriage between the Almoloyas Mixtecs and lords of Cuilápam, not a result of flight from the fierce Cuicatecs.[111]

Still, while the Cuicatecs may have warred with Mixtecs over control of water from the western side of La Cañada, their *cacicazgos*' political influence does not appear to have exceeded the reaches of the canyon. Within the canyon, however, they controlled water and land and would resist any outside interference. The valley floor is semiarid, lacking sufficient rainfall to provide reliable harvests, so irrigation is essential. Whoever controlled water also controlled the regional economy and the destiny of the settlements.[112]

109. Cuicatecs are Otomanguean speakers with distant affinity to Mixtecs and even more distant affinity to Zapotecs. Their *cacicazgos* were independent of those of the Mixtecs and Chochos. Cuicatec settlements are now confined to the steep eastern and western slopes of the Sierra Juárez east of Cuicatlán.

110. Caso in Paddock 1966, 329.

111. Rincón Mautner 2015, 184.

112. This insight was elaborated by Mexican anthropologist Angel Palerm, who refined Karl Wittfogel's theory of the origins of Oriental despotism and applied it to the Mesoamerican region, highlighting the role of hydraulic works on societal development. See Palerm, Maestre Alfonso, and González Jácome 2007, 37.

La Cañada is far smaller in area and contains much less arable land than the Valley of Tehuacán (which includes the Valley of Zapotitlán). The Valley of Tehuacán (depending on the geographer's proposed boundaries) is 170 kilometers long and 40 kilometers wide.[113] La Cañada is only about 20 kilometers north to south and 3–4 kilometers wide at its widest point. Though La Cañada is more tropical than Tehuacán Valley, arable sites, that is, locations for fields suitable for crops, are restricted to four alluvial fans where watercourses enter the Río Grande or the Río de las Vueltas from the east or west: Quiotepec, Cuicatlán, El Chilar, and Santiago Dominguillo. La Cañada's regional strategic location connecting Mexico's two landmasses rendered the canyon more important, and those alluvial fans, all with manageable and deep soils, are fertile and readily amenable to irrigation. All land suitable for crops, up to the last square inch, was apportioned hundreds of years before the arrival of Europeans and provided with complex irrigation systems. The first residents took advantage of these resources, and cultures evolved over a period of at least two millennia. With irrigation, the milpas in La Cañada produce two crops of corn each year, and the valley could easily have supplied corn, beans, and squash for an army of modest size. The low elevation and accessible water of La Cañada are also ideal for harvesting large quantities of wild and cultivated tropical fruits during their season.

Trade between Cuicatlán and the central valleys of Oaxaca is recorded as early as the Perdido phase of Cuicatlán archaeology, beginning roughly 400 BCE. At that time Cuicatecs were already providing fruits and nuts to Oaxaca, perhaps in exchange for Monte Albán pottery. They appear also to have been using obsidian tools imported from central Mexico.[114]

Archaeological reconstruction, however, shows that since precolonial times, Cuicatecs tended to communicate north to Teotitlán and Tehuacán and east to Veracruz, rather than south to Oaxaca.[115] This may have been due to the easier accessibility to the north and east than to the west and south.[116] (The relatively populous Valley of Tehuacán is separated from La Cañada only by a low ridge and a few minor hills, whereas the central valley of Oaxaca is separated from La Cañada by deep canyons and precipitous mountain slopes and was traversed only by trails renowned for their steepness.) The Cuicatecs may also have desired to escape the political dominance of Zapotec and Mixtec *cacicazgos*, preferring the more equal terms of trade they

113. C. E. Smith 1965a, 55. His boundaries are generous, perhaps excessively so.

114. Spencer and Redmond 1997, 600.

115. Hunt 1972, 168.

116. Morante López (2010) documents the large network of roads connecting Tehuacán in all directions. A major road linked La Cañada and the enclave of Tultepec in the lush, tropical lowlands of the Gulf coast, followed by foot and by dugout canoes through the deep canyon of Río Santo Domingo, and by fluvial navigation on the mighty Río Papaloapan.

experienced with the Popolocas of Tehuacán.[117] Residents of La Cañada still travel to Tehuacán for their shopping. The distances between the two destinations are roughly the same, but the drive to Oaxaca involves numerous mountain grades and curves, while much of the drive to Tehuacán is comparatively straight.

What do we know about the Cuicatecs' economic and political role in pre-Columbian Oaxaca? In Eva Hunt's 1972 monograph, she identified thirty-five Cuicatec "population nuclei" that survived out of forty mentioned in documents from 1560.[118] That high percentage indicates a heritage of post-Conquest political stability in the region. In pre-Columbian times, six of these nuclei were to a greater or lesser degree *cacicazgos*, also called *señoríos* in colonial documents.[119] All were included in three *provincias* (probably a Spanish alignment): Cuicatlán, Pápalo-Teutila, and Teotitlán (or Tecomavaca).[120] The latter may have been Cuicatec prior to the Aztec conquest in the late fourteenth or early fifteenth century but is not currently so. Tecomavaca's last Indigenous occupation was probably Náhuatl or Mazatec. In addition, some Cuicatec settlements have survived on the eastern slopes of the Sierra Juárez, notably Peña Verde, which is not mentioned in the Hunt monograph. Hunt lists Atlatlauca (*sic*) as a Cuicatec *cacicazgo*, though today residents proclaim themselves to be Chinantecs.[121] It, Alpizagua (now known as Santiago Dominguillo), and Zoquiápam Boca de los Ríos are located on a lateral drainage of the Río Grande known locally as Cañada Chica. The river that flows by them is called Río de las Vueltas. It enters the Río Grande at El Chilar. "Vueltas" refers to the numerous river crossings that confront the traveler hoping to connect between Tehuacán and Oaxaca. We pored over maps and hiked over a few hills, hoping for a roadway close

117. Popolocas are Otomanguean speakers who occupied much of Tehuacán Valley prior to invasion by Náhuatl speakers. They were responsible for monumental architecture throughout the central and northern parts of the valley.

118. Del Paso y Troncoso 1905. These populated areas are mentioned in a collated list of tribute-paying encomiendas from the mid-sixteenth century.

119. Hunt (1972, 171) lists the nuclei as Atlatlauca (referred to as Atatlahuca, meaning "vermillion water," by local Oaxacans and as San Juan Atatlahuca by INEGI, Mexico's national geographic bureau), Alpizagua (Santiago Dominguillo), and Cuicatlán, all located in the valley bottom, as well as (Concepción) Pápalo, Tepeucila, and Tutepetongo, located in the highlands on the eastern slopes of La Cañada (the western slopes of the Sierra Juárez). Tepeucila and Tutepetongo lie only 10 kilometers or so distant from each other but are separated by a profound canyon that requires more than an hour to traverse.

120. At least four Cuicateco settlements are named Pápalo (Náhuatl for "butterfly"), each use of the name preceded by a saint's name or religious designation. All are located on the western slopes of the Sierra Juárez. Teutila is located above the Río Santo Domingo, well downstream from Quiotepec. It is still regarded as a Cuicatec town.

121. Hunt's language map also indicates that Soquiapa (*sic*) (Zoquiápam Boca de los Ríos) is linguistically Cuicatec. Zoquiápam is an *agencia* of the municipio of Atatlahuca, and its residents also consider it to be a Chinantecan town. We suspect that both pueblos have long been Chinantecan, and that Hunt's map may reflect some historical inaccuracies in colonial Spanish demography.

to the river bottom that would allow us to drive to Atatlahuca upstream from Alpizagua. The steep canyon sides preclude that possibility. We failed.

Each *provincia* contained *cacicazgos* and various towns or hamlets dependent on them. Since rainfall in the bottom of La Cañada averages less than 500 millimeters annually and is inadequate most years for raising corn and beans, Cuicatec *cacicazgos* and *señoríos* were dependent on irrigation. A few settlements at higher elevation could survive on *temporales,* milpas watered by rainfall alone, but these did not support large communities (Peña Verde on the moist eastern slopes of the Sierra Juárez needed no irrigation then or now). The source of water for the *cacicazgos* was not the Río Grande itself but entirely or nearly entirely minor streams originating in the Sierra Juárez or the Mixteca Alta and entering the Río Grande from the east or west. Along lengthy stretches, the river's currents were too swift and erosive, and the gradient of the river generally too steep, for water to be readily diverted into irrigation ditches. The canyon of Río de las Vueltas (Cañada Chica) and Tomellín Canyon were also the locations of important agricultural settlements. Their waters are more manageable than those of the Río Grande.

Since irrigation of crops was essential for survival of Cañada communities, maintenance of a reliable water supply for irrigation was vital for the *cacicazgos.* Skirmishes, battles, and even wars were fought over water. For example, between 1172 CE and 1198 CE Yepaltepec, far upstream on a tributary drainage of the Río Grande, defeated Quiotepec in a decisive battle over ownership of the stream's water. This seminal event remained operative for centuries in determining water allocation. One result was that Quiotepec agreed to pay tribute (taxes) to Yepaltepec for use of the water. The pivotal water battle of the twelfth century was invoked in litigation between the two settlements in the mid-1960s.[122] Struggles over control of water also resulted in the marriage in 1458 of Citlatecutli, the cacique of Pápalo and Tepeucila, to the *cacica* (female cacique) of Quiotepec in an attempt to reconcile a three-centuries-old water feud between the two principalities.[123] The reconciliation, if it occurred, was apparently only temporary. Water battles endure as long as water needs compete.

In fact, records show that intermarriage between *cacicazgos,* even between different ethnic and linguistic groups, frequently took place in the pre-Columbian Cuicatlán region, as elsewhere. These typically cemented power relationships among nobility or overcame otherwise insoluble feuds and divisions, thus solidifying access to irrigation water.[124] The particulars of succession were inscribed in codices, which the *cacicazgos* guarded jealously over the centuries as historical records.

122. Hunt 1972, 198.

123. Hunt 1972, 223. Note that the names are Náhuatl, not Cuicatec.

124. Hunt 1972, 225–31. Parts of some communities (Atatlahuca) were rewarded as encomiendas, and several sugarcane haciendas were founded early in the seventeenth century.

FIGURE 4.10 Peña Verde, a hamlet on the upper eastern slopes of the Sierra Cuicatlán.

For most of the sixteenth century, Spanish administrators found it prudent to leave these political structures and territories largely intact, demanding only that the Cuicatecs pay taxes (instead of tribute). During the seventeenth century, however, pressure from settlers led the Crown to turn over more and more of the most productive land and water rights, including control over the waters of La Cañada and the best delta lands, to Spanish colonists, quickly undermining the authority of traditional caciques, eventually relegating Cuicatecs to the mountain slopes. The exception as of 1972 was Concepción Pápalo, which seems to have retained its pre-Hispanic borders and local control of water and land, though its location is far up the slopes of the sierras, and the slopes within the municipio are steep and require intense management.[125]

Hunt's analysis of irrigation and canal operation demonstrates that organization of the hydraulic system was intertwined with social structure. Pre-Columbian crops included chicozapotes, cotton, corn, beans, squash, and chiles.[126] The ditches also irrigated orchards of coyol palms (*Acrocomia aculeata*), used for oil, fermented beverages, and roofing.[127] Some of these products, especially the chicozapotes and coyol palm products, were traded with Oaxaca, perhaps as tribute, perhaps in exchange for ceramics. Archaeological studies reveal a dramatic increase in production of coyol following the Zapotec conquest of La Cañada, probably an indicator of increased demand from the Zapotec conquerors.[128] Post-Columbian introduction of sugarcane, rice, and mango orchards complicated water allocation, placing new and onerous demands on the Cañada's water supply. Mango and later lime (and other citrus species) and tamarind complemented pre-Hispanic fruit orchards. Rice and sugarcane production, however, are heavy water users and predictably gave rise to large plantations under single ownership. Hacienda owners amassed the political muscle to monopolize irrigation water for their own needs and to bend politicians to their will, thus creating a new class structure. Their fields now rely on pumped groundwater.

Hunt suggests that Cuicatecs were united internally by language and tradition. For the most part, disputes among *cacicazgos* were settled peaceably. She posits Aztec conquest of La Cañada as occurring in 1510.[129] This action would have followed Ahuízotl's campaign against Tehuantepec by fourteen years. Cuicatecs would

125. Hunt 1972, 210, 234.

126. Redmond 1983, 31. The town of Atatlahuca now proclaims itself the center of chicozapote production and is famous for the largest and sweetest fruits. The town plaza seems to support the claim: it is surrounded by lush trees that in season are heavy with the fruits, which are the size of tennis balls. The tree *Manilkara zapota* is also the source of chicle, the raw material for the original chewing gum.

127. Spencer and Redmond 1997, 514.

128. Spencer and Redmond 1997, 515.

129. Hunt 1972, 210, 234.

FIGURE 4.11 Chicozapotes growing in the plaza of Atatlahuca, Cañada Chica, which now claims to produce the best fruits.

FIGURE 4.12 Río Grande and pre-Columbian irrigation *apantle*—a ditch with modern lining. Cuicatlán lies 5 kilometers beyond the bridge.

have been Aztec subjects for little more than a decade before meeting their new conquerors, the Spaniards.

La Cañada in Prehistory

As we have seen, Tehuacán and Cuicatlán Valleys were corridors for soldiers, merchants, and migrants. But they were also home to developed civilizations, both Indigenous and occupied. No single points along the route matched the prominence of the Popolocan structures on La Mesa (Tehuacán Viejo), the strategic significance of Cerro de Quiotepec, or the economic reach of the Purrón Dam. While only some 40 kilometers separated these structures, their cultural and social roles differed dramatically. For southward-bound travelers prior to the Common Era, their greeting as they crossed into La Cañada was a defensive fortress.

Cerro Quiotepec is a steep, though not massive, hill overlooking the confluence of the Río Salado, which drains the Valley of Tehuacán, and the Río Grande, which drains La Cañada and upstream penetrates deeply into central Oaxaca.[130] Downstream from the confluence, the river, now renamed the Santo Domingo, enters a narrow gorge at the base of steep mountains. That steep canyon is the only gap through the long mountain chain formed by the Sierra Zongolica and the Sierra Mazateca on the north and the Sierra Juárez on the south. In the early Quaternary, some 2.5 MYA, the large lake contained in the valleys drained through this gap, dramatically changing the character of the whole region. The cordillera extends otherwise unbroken from Mexico's central plateau to the Isthmus of Tehuantepec and has formed a permanent obstacle to east–west traffic. The mountain mass has produced rain forests to the east and a rain shadow to the west, creating an arid zone in the valleys.

Atop the cerro—a natural bulwark—sits a complex pre-Columbian fortress with a commanding presence over the confluence of the rivers and the approaches from every cardinal point. From the fortress and outlying watchtowers, guards could observe the passage of all but the smallest groups of passersby along the valley floors and from the mountains to the east and west. The Valley of Tehuacán tapers to a narrow gorge at its southern extremes, but a benign pass of such low relief that a traveler hardly recognizes it as such allows easy crossing into La Cañada. From there,

130. Quiotepec is a Náhuatl name: *quiotl* = agave stalk; *-tepetl* = hill. Much of our archaeological information from La Cañada is derived from the pioneering investigations in La Cañada carried out by Charles Spencer and Elsa Redmond beginning in 1977 and summarized in their extensive report from 1997. A key source for historical information on La Cañada and environs is Eva Hunt's study of Cuicatec *cacicazgos* based on archival sources (Hunt 1972).

the terrain slopes gradually upward to the south via a series of deltas that form a staircase.

The valley of La Cañada follows the Río Grande upstream from Cerro Quiotepec some 20 kilometers, where it splits near the town of El Chilar into two forks. On the east fork, the Río Grande proper encounters encroaching mountainsides that squeeze the waters into a narrow and profound canyon that trends east-southeast and is without substantial agricultural potential. A grade-level diversion structure some 2 kilometers upstream from the junction now diverts water into a large canal, which supplies orchards on the right (east) bank of the Río Grande well north, nearly to Cuicatlán. Otherwise, the Río Grande runs free some 40 kilometers upstream, where Highway 175 crosses it on a bridge not far south of Ixtlán de Juárez.

The western, smaller fork is known as Río de las Vueltas, while its canyon is called Cañada Chica.[131] This river, named for its interminable curves and bends, which require numerous crossings for the foot traveler, has ample alluvial fans with deep, fertile soil sufficient for supporting intensive agriculture. The river originates nearly due south of La Cañada in mountainous country only a few kilometers northwest of the city of Oaxaca. It is separated from the Río Grande by the large and lofty mountain range known as the Sierra Monteflor. Mexican botanists lament the near absence of botanical explorations in this steep and forbidding range.[132]

The town of San José del Chilar sits on an alluvial fan where the Río de las Vueltas enters the Río Grande from the south. Upstream in the Cañada Chica, along the route most probably taken by Aztecs and colonial Spaniards, lies the town of Alpizagua (Santiago Dominguillo).[133] Farther upstream, some 15 kilometers from

131. The Capuchin padre Fray Francisco de Ajofrín traveled south on the Camino Real from San Juan de los Cues in the Tehuacán Valley, 6 leagues (roughly 20 mi.) across the Río Salado and the Río Grande (also known as the Río Quiotepec), to Quiotepec in March 1766 (Ajofrín 1959, 47). Ajofrín's account lists the itinerary and the distances for the journey (Spencer and Redmond 1997, 621). From Dominguillo, Ajofrín continued southward up the Rio de las Vueltas to Atatlahuca, some 5 leagues distant, and on to Jayacatlán, another 5 leagues away, where he spent the night. The next day, Ajofrín climbed the mountains about 6 leagues on the Camino Real to San Juan del Rey (San Juan del Estado), having crossed the Río de las Vueltas a total of sixty times. San Juan del Estado was the first Zapotec-speaking town he encountered. From here, Ajofrín left the Camino Real to Oaxaca and proceeded 4 leagues to "Guaxolotitlán" (Huitzo) in the Valley of Oaxaca (Ajofrín 1959, 83–85). Ajofrín noted that another route could be taken to avoid all the "vueltas," especially during the rainy season, when the route up the Río de las Vueltas was impassable. "This alternative route from Don Dominguillo [Santiago Dominguillo] involved the 4-league-long ascent on firmer terrain to San Pablo Cotahuistla, and from there 7 leagues to San Francisco de Huiso and 1 league to Guaxolotitlán, making it a shorter and safer trip to the Oaxaca Valley" (Ajofrín 1959, 84). On Ajofrín's return from Oaxaca to Mexico, he journeyed through the Mixteca, a route that avoided the mighty rivers of the Cañada route, but which proved to be 11 leagues longer and fraught with smaller rivers to cross. Ajofrín resolved never to take this route to Oaxaca again, and he advised others to follow the Camino Real through La Cañada.

132. Patricia Dávila and Oswaldo Téllez, personal communication, 2014.

133. Hunt 1972, 169.

FIGURE 4.13 Cañada Chica south of Atatlahuca. The mountain range to the left is the Sierra Monteflor.

Alpizagua, is Zoquiápam Boca del los Ríos, and 3 kilometers more upstream, Atatlahuca. Both towns are located around alluvial fans deposited from side canyons or tributaries.[134] An additional 15 kilometers upstream, through a land of terraced milpas below and rich tropical deciduous forest above, lies San Juan Bautista Jayacatlán, also situated on an alluvial fan, planted with modest orchards of chicozapotes and, at their upper elevational limit, a host of milpas and maguey fields, as well as a few coffee trees.[135] Upstream from Jayacatlán the stream gradient increases as the canyon narrows. Oak- and pine-covered hills close in on the drainage. Another 15 kilometers upstream, over a steep but friendly grade through pleasant, well-watered forest, Cañada Chica ends at a low pass. Three kilometers below, the village of San Juan del Estado is nestled at the head of a temperate and well-watered valley, which connects easily and directly with the verdant and fertile Valley of Etla, immediately north of

134. La Cañada usually refers to the stretch of canyon between El Chilar, where the Río de las Vueltas (or the Cañada Chica) joins the Río Grande, as far north as Quiotepec, where the Río Grande joins the Río Salado to form the Río Santo Domingo.

135. In this area, coffee trees require sufficient rainfall, warm (but not hot) temperatures, and some shade from trees overhead. In the Cuicatlán drainage, only Jayacatlán possesses the requisite conditions.

FIGURE 4.14 San Juan Bautista Atatlahuca, Río de las Vueltas, an ancient Chinantecan town.

FIGURE 4.15 Jayacatlán, Río de las Vueltas. From here, it is roughly 20 kilometers to San Juan del Estado and the Valley of Etla in Oaxaca's central valleys.

Oaxaca City. The remaining towns of La Cañada, separated by a steep mountain range from the Cañada Chica, are located along drainages on mountain slopes or are situated atop hanging gardens far above the canyon bottom of the Río Grande.

La Cañada experiences the hottest temperatures in the state of Oaxaca. Temperatures of 40°C (104°F) are not unusual, especially during April and May. French adventurer and educator Mathieu de Fossey visited La Cañada in 1837, en route to Oaxaca, fleeing from the fetid air and foul streets of Mexico City. He put up for a few days in the canyon at the hacienda called Guendulaín, a sugarcane plantation near the former railroad settlement of Valerio Trujano. Although he was fascinated by sugarcane production, he quickly discovered the painful reality of summer in La Cañada: "It did not take me long to tire of life in Guendulaín. As this valley is surrounded by high mountains, you never experience a cooling wind, and the heat is oppressive in the months of April and May. Day and night you are drenched with a boiling sweat, and even though you lie at night with no cover at all, you cannot detect the slightest breath of a breeze."[136] The torrid heat, limited agricultural production, broken terrain, and the absence of ample building sites explain the absence of cities. All lands of gentle or no slope were dedicated to crops, leaving hillsides the only option for constructing houses and administrative and religious structures.

Just why the valleys between the cities of Tehuacán and Oaxaca are not as densely populated as the Basin of Mexico to the north and the valleys of Oaxaca to the south is best explained in Fossey's words: "Arriving finally at the gentle slopes at San Juan del Estado [at the southern terminus of the Río de las Vueltas], one comes upon the first of the three valleys of Oaxaca, called the Valley of Etla. Just six leagues farther on it joins with the other two valleys, forming with them the junction of the three-branches where the city of Oaxaca is located. There one enjoys perpetual spring, which gives rise to a most unusual and varied vegetation growing all in one zone."[137]

Archaeologists Charles Spencer and Elsa Redmond endured the full daytime blasts of several Cañada summers. Their findings led them to argue forcefully that the Zapotecs had invaded La Cañada and expanded the fortress of Quiotepec in response to a growing threat from Teotihuacan (and, apparently, a lack of suitable defensive zeal from the local inhabitants).[138] The cerro sits atop the lowest and hottest point in the entire intervalley stretch, so assignment of Zapotec troops to the fortress was probably not a reward for exemplary behavior. But stock it with troops the lords of Monte Albán did, quite possibly until trade waned, the threat from Teotihuacan faded, or manning the fort became too expensive. They slowly abandoned the site beginning sometime after 200 CE, a time when Monte Albán's military

136. Fossey, cited in Ramírez 1958, 142–43.

137. Ramírez 1958, 143.

138. Spencer and Redmond 1997, 609.

presence began to recede throughout its region of influence, and the military power of the Zapotec state declined or was directed elsewhere. (Even so, Monte Albán's assemblage of monuments continued to expand for several hundred years.) After roughly 600 CE Monte Albán was no longer the dominant force in the central valleys of Oaxaca, though it remained a large population center.[139]

At the time of construction of the great defensive structure atop Cerro Quiotepec, Zapotecs held sway over most of upland Oaxaca, if not directly then at least by dint of the threat of military action. Trade between Cuicatlán and Oaxaca central valleys is recorded as early as the Perdido phase of Cuicatlán archaeology, beginning roughly 400 BCE.[140] Cuicatecs were then providing cotton, fruits, and nuts to Oaxaca, perhaps in exchange for Monte Albán pottery, which also appears on the site. The Cuicatecs appear also to have been using obsidian tools imported from central Mexico.[141] Long before that, traders undoubtedly frequented the route through the valleys.

During the same time, midway through the first millennium BCE, the Mixtecs in the mountains and canyons that rise west of La Cañada had developed a culture contemporaneous with Monte Albán, with advanced monumental construction, religious and administrative centers, and sophisticated artisanal accomplishments. They never achieved Monte Albán's level of military dominance, however, and were never represented by a strong urban center.

Mixtecs emerged from relative political obscurity as Monte Albán languished during the transition from the Postclassic (Monte Albán IV) into the Late Postclassic (Monte Albán V). The former refers to the time of the disintegration of the Zapotec state, while the latter is related to the penetration of the Mixtecs pushing south through the adjoining Etla territory into the central valleys of Oaxaca. Using codices, for which the Mixtecs were famous, Alfonso Caso constructed a genealogical history of Mixtec ruling families extending back to 692 CE.[142] He found that at least some of their traditions relate their rise in power in the La Cañada region. These ancient narratives were painted on amate, a paper produced from fig bark. Others were painted on cotton cloth, agave fibers, or deerskin, usually sized with lime, which provides a white background. In rare cases, like Codex Laud (the book of death from central Mexico), codices were bound in jaguar skin. The codices were long sheets typically folded, accordion style, recording family trees, important historical or celestial events, and successions of royalty. These were made by scribes (*tlacuilos* in Náhuatl) following the orders of high-ranking officers or priests.

139. Balkansky 1998, 475.

140. Spencer and Redmond 1997, 602.

141. Spencer and Redmond 1997, 600.

142. Spores 1967, 59.

Regarding the Mixtec codices, Jiménez Moreno notes that

> the Mixtec codices . . . register a series of wars that these inhabitants of [now] poor eroded lands started to wage in A.D. 720, the year, to all appearances, in which the principality of Tilantongo was founded. Father Antonio de los Reyes, in the beginning of his *Arte en lengua Mixteca* (1593), refers to traditions according to which this dynasty originally is from Apoala, whence a few lords, certainly foreigners, embarked on the domination of the "true Mixtecs." Since the cradle of the dynasties that ruled the Mixtecs was situated at the source of a river that flows down to the Tomellín Canyon [La Cañada], where Cuicatlán and Quiotépec [*sic*] are located, we can conjecture that the founders of these dynasties were perhaps Cuicatecs, and therefore related linguistically to the people whom they enslaved. Cuicatlán and Quiotépec [*sic*] at this time must have been a cultural crossroads, since in Cuicatlán a yoke of Tajín [Totonacan—from Puebla, San Luis Potosí, Tamaulipas, and Veracruz in northeastern central Mexico] affiliation was found.[143]

Apoala (now referred to as Santiago Apoala) is an isolated Mixtec agrarian community of fewer than five hundred inhabitants. It lies in a small verdant bowl-shaped valley at an elevation of roughly 1,900 meters, in the upper reaches of a canyon that empties into La Cañada from the west near Cuicatlán, immediately north of the mouth of Tomellín Canyon, a major tributary of the Río Grande. Massive limestone cliffs border the Apoala valley on three sides. On the Cañada (eastern) side, a drop of more than 100 meters creates a succession of waterfalls, the lowest one more than 30 meters high. It forms a pool contained in a collapsed karstic formation, marking the downstream limit of arable flatlands within the tiny basin. This fortuitous geological feature plugged the canyon, trapping sediments for eons and ultimately forming a sizable terrace or hanging garden. The accumulated sediments, now many meters deep, provide flat fields, easily irrigated from the faithful stream that runs through the valley, and fertile soils that have made it prosperous, a bucolic agrarian enclave easily defended in all directions.

Pre-Columbian painted figures referring to Apoala ("The Apoala Toponym") appear in what are known as the Codex Nuttall and the Codex Vindobonensis. Both codices established Apoala as the source of the Mixtec royal lines. (Other traditions trace Mixtec origins to a different locale, such as Sosola, some 30 kilometers east, also in La Cañada drainage.[144]) The narratives portray the royal families as founded by alien conquerors and identify Apoala as the birthplace of the new race of Mixtec

143. Jiménez Moreno 1966, 61. Moreno suggests that the "foreigners" were most likely from Teotihuacan but does not offer dates of the occupation.

144. Burgoa (1674) 1934, 274, 275; Nuttall 1974, 36; R. Williams 2009, 125; Furst 1978, 37.

rulers, who would dominate Mixtec politics from the downfall of the Zapotecs until the Conquest. The codices date Mixtec genealogy from no earlier than about 937 CE, so at least two millennia of Mixtecs had already transpired prior to the emergence of Apoala as the source of the Mixtec royalty.[145]

The Codex Nuttall depicts the waterfall for which Apoala is renowned as part of the creation of the world. The Codex Vindobonensis depicts the tree of Apoala, royalty emerging from a cleft in the sacred tree, clearly identified with Apoala. The tree, growing in a place called Cahuacandehvui ("the rock outcrop where the heavens rest"), produced the first four beings after being impregnated by a man or a god.[146] So, while the bulk of the Mixteca Alta lies beyond the drainage of La Cañada and the Valley of Tehuacán, its most notable place of mythical Mixtec origin lies well within the valleys. Although the origin for the new Mixtec nobility was clearly traced to Apoala by Caso, three places are associated with the birth of modern Mixtecs: Apoala, as the place of the sacred land where the nobility was born and where the new order was established; Monte Albán, appearing in the Mixtec codices as the sacred place of ancient relics and divinity; and Tilantongo, near Nochixtlán, as the place of the new Mixtec rulers. Tilantongo is located near the ancient site of Monte Negro, which Caso demonstrated to be among the oldest Mixtec settlements in Oaxaca, along with Yucuita and Yucuñudahui.[147]

Thirty kilometers north of Apoala, the Mixtec village of San Pedro Nodón, downstream from Santa María Ixcatlán and of great antiquity, is perched on the edge of La Cañada. It sits 1,000 meters higher in elevation than the town of Cuicatlán, whose market Nodón and Ixcatlán residents routinely visit. This geographic closeness of Mixtecs to La Cañada, however, has not yet translated into evidence of cultural or social penetration of the canyon bottom. Mixtec influence is equivocal as well in archaeological sites in the valley bottoms, along the Río Grande and the Río de las Vueltas: no clear signs of domination of Mixtec *cacicazgos* in La Cañada have turned up, nor have significant amounts of Mixtec artifacts. This suggests that Cuicatecs controlled La Cañada while Mixtecs controlled the western slopes and canyons, as they had for millennia. Zapotecs kept Mixtecs at arm's length during their five hundred-plus year occupation. Hunt believes Mazatecs (from the east) dominated Quiotepec and nearby Tecomavaca in the twelfth century at the very time the Quiotepecs were doing battle with their upstream enemies from Yepaltepec.[148] If she is correct, the battle over control of water would have been waged between

145. R. Williams 2009, 40.

146. Fr. Antonio de los Reyes in his *Arte en lengua mixteca*, cited by Jiménez Moreno (1966, 61).

147. Jansen 1998, 114–15. While the significance of Apoala and Tilantongo is well established among archaeologists, the connection with Monte Albán remains controversial.

148. Hunt 1972, 211, 233.

FIGURE 4.16 Apoala, Mixteca Alta, in La Cañada drainage. Mixteco legendry makes the village the birthplace of Mixtec nobility.

FIGURE 4.17 Waterfall below Apoala. It drains the small Apoala basin.

FIGURE 4.18 The tree of Apoala from the Codex Vindobonensis. It shows the first of Mixtec nobles emerging from the earth through a cleft in the tree. Courtesy of the British Museum.

FIGURE 4.19 San Pedro Nodón, an ancient Mixtec village between Santa María Ixcatlán and the Valley of Cuicatlán. The terraces, though ancient, are still productive.

FIGURE 4.20 View to the east from San Pedro Nodón, overlooking La Cañada. White gashes are recent erosion from a microburst thunderstorm. The lower canyon provides habitat for endangered military macaws.

FIGURE 4.21 Chinantecan father and son, Atatlahuca, Río de las Vueltas. Chinantecs appear to have lived here for at least one millennium.

peoples of different ethnicity (upstream Cuicatecs vs. downstream Mazatecs) rather than through an internecine struggle among Cuicatecs. Furthermore, Cuicatecs and Mixtecs are close linguistically, so their developmental origins may have taken place close to each other.

By the time of the Conquest, the fortress of Quiotepec had been largely abandoned.[149] Quiotepec and Tecomavaca town sites were still occupied by Mazatecs, while Cuicatecs were confined primarily to settlements on the western flank of the Sierra Juárez (which of the two groups exercised control of the bottomlands of La Cañada is unclear). It was not until the eighteenth century that Cuicatecs came once again to dominate the fertile alluvial fans of La Cañada. Today Atatlahuca, an agricultural town in Cañada Chica, and its *agencia,* the nearby town of Zoquiápam, are Chinantecan, not Cuicatec. Whether this is a recent or ancient ethnic affiliation we cannot determine.

A Brief Glimpse of the Archaeology of the Valley of Tehuacán

The Zapotec soldiers, engineers, and architects from Monte Albán were latecomers onto the La Cañada scene. The paths through the valleys were already well worn when the Monte Albán conquerors went to work at Quiotepec. Archaeological studies at the Purrón Dam site (also commonly referred to as Mequitongo Dam),

149. Hunt 1972, 213.

situated between Coxcatlán, Puebla, and Teotitlán de Flores Magón, Oaxaca, some 30 kilometers north of Quiotepec, reveal a sophisticated dam structure plus a network of canals, diversion channels, and weirs dating back as far as 750 BCE. The largest of the dams at the site ultimately reached 400 meters in length, 106 meters in width, and 24 meters in height.[150] In its final stage of development, early in the second millennium of the Common Era, the dam served as a base for ceremonial and religious structures rather than as a water storage and management facility. Still, the dates during which the dam operated as an irrigation structure underscore the active agricultural production of the valleys centuries prior to the Zapotec occupation of Quiotepec.

At roughly the same time as the dam had reached its full operational status, around the beginning of the Common Era, several locations in the Valley of Tehuacán became centers of obsidian warehousing and, quite possibly, obsidian technology and production. These way stations in the valley continued to act in that capacity through Middle Classic times, that is, for at least five centuries.[151] They were not urban centers. They were probably no more than small single-industry settlements, where middlemen and craftsmen received shipments of raw obsidian and produced finished obsidian blades and tools. They provided merchant services as well as repair of old or broken tools along with production of new ones. Craftsmen or their bosses probably traded finished obsidian implements with other polities for high-quality goods in which they, in turn, specialized. Some of the workshop sites occupied hilltops, selected with defense in mind. In addition, some sites in the vicinity of Tehuacán were also trade or distribution destinations for the celebrated Thin Orange pottery, apparently produced during the Classic period by Popolocas in Tepexi de Rodríguez, a Popolocan town northwest of Tehuacán. It was a highly valued ceramic throughout central Mexico at the time. And thus, these modest sites were nodes for international trade.

Popolocas were active in the Late Classic and Postclassic periods as well, constructing monumental architecture at the sites of La Mesa (Tehuacán Viejo) and Cerro Cuthá. Whether these accomplishments were of purely Popolocan origin or represent an ethnically diverse culture is unclear, but the area of southern Puebla occupied by Popolocas demonstrated grand civic and ceremonial centers, the accomplishments of a complex culture.

The role of the Valley of Tehuacán in human history is much more than mere trade routes, however. In the early 1960s Richard MacNeish spearheaded the Tehuacán Project, a multidisciplinary initiative, to discover "processes and causes leading

150. Woodbury and Neely 1972, 81–153; Aiuvalasit, Neely, and Bateman 2010.

151. Drennan, Fitzgibbons, and Dehn 2000, 176–88.

FIGURE 4.22 Coxcatlán Cave, where scientists headed by Richard MacNeish established the great antiquity of corn and cultures of the Valley of Tehuacán.

to the development of primary civilization."[152] MacNeish, who had for some decades studied the origins of maize, believed he had discovered remnants of the oldest domesticated plants, roughly 7,600 years old (5600 BCE), in a cave near Coxcatlán. He believed that this proved the Valley of Tehuacán was the location of the original domestication of maize. Subsequent evaluations of the Tehuacán Project specimens based on a more sophisticated analysis, however, reveal a much more recent date of the remnant plant parts. Most authorities now agree that maize cobs from Guilá Naquitz Cave in Oaxaca, east of Oaxaca City, are considerably older. Still, MacNeish's discoveries were revolutionary, and he pinpointed the Valley of Tehuacán as a cornerstone in the early domestication of corn and in the selection of strains most responsive to irrigation and tolerant of arid conditions.[153] Early Mesoamericans experimented with corn throughout Mesoamerica, from Tamaulipas in the far northeast to the Petén in Guatemala, and from the Gulf of Mexico coast to the Pacific slopes location of Manantlán, Jalisco, where perennial teosinte (*Zea diploperennis*), a wild relative of corn, was discovered growing in the 1980s. If the earliest domestication site was in the eastern Balsas Valley (Mexico's largest Pacific

152. MacNeish 1967, 14.

153. Long and Fritz 2001; Piperno and Flannery 2001, 2101–3.

drainage) as some researchers now suggest, it lies near the Continental Divide, not far from the westernmost reaches of Zapotitlán Valley.[154] Ironically, perhaps, a team of scientists published in 2014 their findings that Tehuacán Valley was the site of the domestication of chiles.[155] The unending search continues, as it will indefinitely, for the oldest site bearing the remains of corn and the accompanying bragging rights.

MacNeish also established without question the great antiquity of human occupation of the Valley of Tehuacán. As a result of the research he and his associates carried out, records of human habitation extend back to at least ten thousand years in the Tehuacán Valley. We can infer from this lengthy human history and the existence of botanically rich and diverse locations in all four directions that peoples moved back and forth through the valley and into La Cañada as well for many millennia. MacNeish assembled a multidisciplinary group of scientists to trace the development of agriculture as it paralleled the rise of civilizations. Among their many findings was that the aforementioned obsidian warehousing was a late-arriving activity.

Despite the abundance of pre-Columbian sites, the archaeological record of the Valley of Tehuacán is comparatively slim. Publicly accessible sites are limited. At the time of the Tehuacán Project, most archaeologists shared the view, expressed by John Paddock, that no pre-Columbian urban center existed in the Valley of Tehuacán due to its ecologically impoverished countryside.[156] He apparently believed that the valley was incapable of supporting a major monumental center. Subsequent studies reveal an abundance of rich sites, but because of Mexico's bounty of sensational archaeological sites elsewhere, archaeologists had until recently paid disproportionately little attention to the valley, and funding for excavation and restoration had been less accessible. The Tehuacán Project focused on the origins of agriculture and not on archaeological sites per se.

And yet, as Robert Chadwick and MacNeish believed they had discovered, records of pre-Columbian peoples of the Valley of Tehuacán turned up in a collection of Indigenous manuscripts known as the Codex Borgia. They were assembled prior to or slightly after the Conquest and appear to describe culture in the Valley of Tehuacán of the Venta Salada phase, 750–1540 CE (roughly equivalent to the Postclassic), perhaps significantly earlier. Much of the portrayal seems to refer to the quasi-independent principality of Teotitlán (de Flores Magón) in the extreme southern part of the Valley of Tehuacán.[157] If this hypothesis is correct, Mesoamerican cultural status in the valleys paralleled that to the north (Teotihuacan, Tula, Xochicalco) and to the south (Monte Albán, Mitla) in terms of sophistication, cultural practices, and

154. Piperno et al. 2009.

155. Kraft et al. 2014.

156. Paddock 1966, 242.

157. Chadwick and MacNeish 1967, 114–31.

FIGURE 4.23 View from Coxcatlán Cave to the west, including the Sierra Mixteca.

monumental architecture. The Popolocan monumental buildings of Cerro Cuthá and the extensive site of La Mesa (Tehuacán Viejo), near San Diego Chalma on the outskirts of present-day Tehuacán, are locations of considerable pre-Conquest antiquity. Both sites demonstrate monumental architectural accomplishments and date at the very latest from the Late Classic and Postclassic periods (see chapter 6). Aztec tradition states that Moctezuma I conquered the Popolocas in 1456 CE and established them as a client state of Tenochtitlán, but the traditional archival sources yield little information about the Popolocan building sites and their functions at the time of Aztec conquest.[158]

The Popolocas occupy a long period of the Valley of Tehuacán's history. The construction of Presa Purrón through all its phases demanded skilled engineering and irrigation works, as well as agricultural sophistication well back into the Formative period. The ruined dam lies well within the sphere of Popolocan influence. This suggests a deep history involving those peoples immediately north of the dam.

In this summary, we have attempted to demonstrate that archaeologically and culturally, Cuicatlán and Tehuacán Valleys have a history as varied and rich as their more celebrated natural history. The valleys have witnessed three millennia of cycles of conquest, shifting power and trade relations, and the rise and fall of civilizations. They reveal precocious engineering developments, like damming Lencho Diego arroyo and creating a citadel atop Cerro Quiotepec, as well as scientific and technological developments, like the production of fine ceramics and obsidian instruments. Above all, they showcase the development of maize as a major crop in its hundreds of varieties—the most highly modified (agronomically improved) grain in the world. Corn and its ongoing evolution demonstrate human accomplishments that underline the legitimacy of the valleys' international reputation. Their merit stands out even more when framed by the rich tapestry of geological, biological, and ecological marvels that the valleys display, the same as those that attracted and sustained ancient peoples and fascinated later explorers, including contemporary ones. Before we present detailed accounts of archaeological sites, we present an introduction to the native peoples of the valleys, those whose ancestors shaped the valleys' future.

158. INAH, "Tehuacán el Viejo," accessed October 23, 2022, https://lugares.inah.gob.mx/es/zonas-arqueologicas/zonas/14643-tehuac%C3%A1n-el-viejo.html.

CHAPTER 5

Cultures and Languages of the Valleys of Cuicatlán and Tehuacán

The valleys have been crossroads for Indigenous American cultures for millennia. During the last century, the dominant language became overwhelmingly Spanish, and the culture of the urbanized areas dominantly mestizo. The city of Tehuacán is now hardly distinguishable from any other industrialized Mexican city.[1] In the numerous small cities and towns outside Tehuacán, however, strong remnants of Indigenous tongues and cultures remain, and their locations provide revealing glimpses into pre-Columbian settlement and trade patterns, as well as the effects of now-ancient cultural expansion, conflicts, and wars.

Indigenous languages still spoken in the combined valleys include Chinantec, Chocho, Cuicatec, Ixcatec, Mazatec, Mixtec, Náhuatl, and Popoloca. All languages except Náhuatl belong to the eastern branch of the Otomanguean family of languages, a broadly diverse collection of tongues from the south of Mexico. Náhuatl is Uto-Aztecan, which includes many languages from the western United States, including Comanche, Hopi, O'odham, Shoshone, and Yaqui. Náhuatl speakers appeared in the region of the valleys from the language family's origins in central or northwestern Mexico as a result first of waves of immigration and, much later, military campaigns of conquest. Chinantec, Mazatec, and Mixtec, languages widely spoken in Oaxaca, are only marginally represented in La Cañada. In contrast, the town of Santa María Ixcatlán is home to *all* the remaining speakers of Ixcatec. Most surviving Chocho speakers live in the environs of Coixtlahuaca. Zapotec is also an Otomanguean language, but at present it has no speakers represented in the valleys

1. Tehuacán's central produce and food markets, however, retain an Indigenous appearance and are frequented by Nahua vendors and customers. This linguistic anomaly is also true of San Juan Bautista Cuicatlán and of Teotitlán de Flores Magón, both mestizo towns but with markets that are heavily Indigenous.

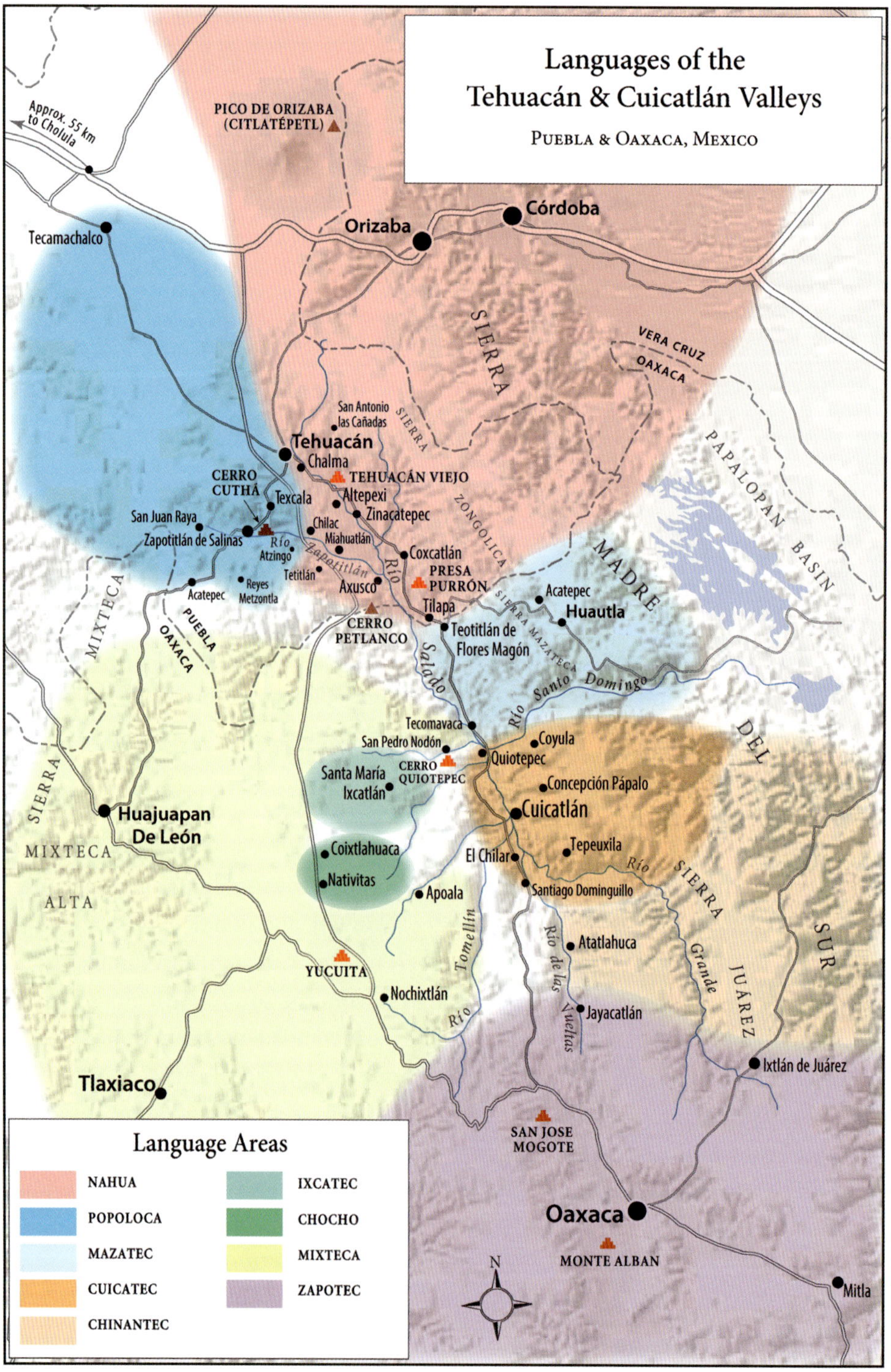

MAP 2 Map of languages of the Tehuacán and Cuicatlán Valleys. Map by Paul Mirocha.

per se. It is still spoken widely, however, in towns and villages of the upper reaches of the Río Grande drainage.

Otomanguean languages are tonal, like Chinese Mandarin and African Bantu, in that speakers determine meaning through changes in the pitch, which they impart in the spoken word.[2] This tonality was alien to Spanish speakers, especially to priests, several of whom had managed without too much difficulty to learn Náhuatl, which is nontonal. Tonal languages, however, were previously unknown to the clergy, and Dominicans, who presided over the evangelization of Mixtecs, reported that they experienced great frustration in learning to speak or to understand the languages, needing decades to master Mixtec, as opposed to a few years for Náhuatl. This linguistic impediment slowed their evangelization efforts, to their frustration. Fray Francisco de Burgoa, for example, described the Mixtec language of Yanhuitlán in 1674 as "a tongue barbarian, inappropriate, and very difficult."[3]

Within the Otomanguean family, Chocho, Ixcatec, Mazatec, and Popoloca exhibit strong affinities. Within this grouping of four languages, Chocho and Popoloca are especially closely related, as are Ixcatec and Mazatec. Cuicatec and Mixtec show a similarly close relationship. Cuicatec, Chinantec, and Mazatec are only distantly connected, though all are Otomanguean, and the regions where they are spoken occupy conterminous territory in northern Oaxaca, primarily east of the Río Grande. As the map demonstrates, the relative geographic positions—the location of neighboring ethnic groups and their boundaries—casts long shadows into history and prehistory. Linguistic borders and linguistic remnants are usually markers of historical struggles or wars, advances and retreats, settlements, and abandonments.

We must bear in mind that the names for various peoples used here for linguistic groups reflect Aztec nomenclature and, except for Nahuas, are not the terms the speakers use to identify themselves. Dominant cultures often apply their own labels to vanquished peoples. Aztec imperialism retains its vast legacy five hundred years after the fall of Tenochtitlán.

Chinantecs. Chinantecs occupy much of the eastern slopes and foothills of the Sierra Juárez of Oaxaca. Linguists identify as many as fourteen different languages within the group.[4] Archaeological investigations reveal little in the way of Chinante-

2. Nicholas A. Hopkins (1984) estimates that the first diversification of this group of languages had begun by 4400 BP, and that the nine branches of the Otomanguean family were already established by 1500 BCE. He also argues that some of this linguistic differentiation took place in the Valley of Tehuacán.

3. Burgoa (1674) 1934, 283; Terraciano 2002, 67–72.

4. Lewis, Simons, and Fanning 2013. Since the evolution of mutually unintelligible languages from a proto language requires long periods, the presence of so many languages implies a lengthy tenure in the Chinantecs' homelands.

can influence in La Cañada, yet according to residents, the agricultural towns of San Juan Bautista Atatlahuca and its *agencia* Zoquiápam Boca de los Ríos, both located in the Cañada Chica, are Chinantecan towns. Eva Hunt considered Atatlahuca to have been Cuicatec in pre-Columbian times, but today there appear to be no native speakers of that language in the municipio.[5] During the Late Postclassic period of Mesoamerican prehistory, between roughly 900 CE and the arrival of Europeans, Atatlahuca figured prominently in the political life of La Cañada. Upstream from Cuicatlán in the drainage of the Río Grande, but outside La Cañada, the settlements of San Juan Quiotepec (not to be confused with Santiago Quiotepec and the nearby fortress far downstream) and Yolox are still home to many Chinantec speakers. Hunt suggests that these towns were also Chinantecan in pre-Colombian times but were allied with Atatlahuca, which she believed was Cuicatec at that time. Bas Van Doesburg identifies them as Chinantecan.[6] Indian national terrain boundaries were in flux throughout that period. But Chinantecs were mainly confined to higher elevation or wetter sites.

Chinantec is only distantly related to other Eastern Otomanguean languages in the region. It may be more closely aligned with the Western Otomanguean languages, most of which are spoken well to the north and include Otomí, a language spoken by a quarter of a million people in the central Mexico highlands north and west of Mexico City.

Despite Chinantecs' widespread distribution, published studies of Chinantecan archaeology and history are fragmentary and elusive. In pre-Columbian times Chinantecs produced fragile pottery, some of it known for its use of yellow and blue, others a characteristic gray. Archaeologists also discovered finely worked gold in some Chinantecan tombs.[7]

Chochos. Chocho, also known as Chocholtec (the Aztec term) and Ñgigua (their term for themselves), is an Otomanguean language within the branch of the closely related Popolocan language. Chocho speakers have dwindled drastically since the mid-twentieth century and now number fewer than two hundred, according to residents of the few remaining Chocho towns. They appear to live in the villages of Santa María Nativitas and San Pedro Monte Verde, both *agencias* of Coixtlahuaca, a region that has experienced significant outmigration in recent decades. A few speakers may remain in the town of Tepelmeme to the north, which lies in the drainage of the Valley of Tehuacán. Coixtlahuaca, regarded as a Chocho region at the time of the Aztec punitive expedition of 1458, was apparently a mixed Mixtec-Chocho town,

5. Hunt 1972, 173–75.

6. Van Doesburg 2001, 405, map notation.

7. Paddock 1966, 227.

but residents of Coixtlahuaca now deny the existence there of speakers other than Spanish and English.[8] They view nearby Nativitas as the Chocho town of record.

Colonial records report that Coixtlahuaca was home to Náhuatl speakers at the time of Conquest, a heritage of the Aztec rampage in 1458. After the Conquest, a 1580 document from Coxcatlán in the southern valley of Tehuacán states of the inhabitants that "among them they are three, which are Mexicana [Náhuatl], which is the main language and widely spoken, and Chochona and Mazateca, languages that are obscure and bad-sounding to our ears."[9] Castillo Tejero believes that Coixtlahuaca constituted the southernmost Popolocan (not Chocho) señorío during the fourteenth century, probably more a reflection of outsiders' difficulty in distinguishing Chochos from Popolocas than demographic reality.[10] Rincón Mautner believes that Chochos split from Popolocas around 1200 CE and moved to Coixtlahuaca from the Zapotitlán Valley.[11] Kevin Terraciano believes that in the seventeenth century, Chochos constituted the clear majority of Coixtlahuaca.[12] Paredes Colín reports that the town (now hamlet) of San Luis Chochos in the northern portion of the Valley of Tehuacán requested a change of name for their village since they considered the name "Chochos" to be demeaning.[13]

Coixtlahuaca is the site of a colossal ex-convent, constructed in the mid- to late sixteenth century by Chochos and (we must presume) Mixtecs under the direction of Dominican priests.[14] The priests may well have decided that only a huge building would sufficiently impress the Indigenous folk of the need to become Christians inwardly as well as outwardly, a daunting task. The cost of the building must have been staggering and possible only in a region sufficiently well populated and affluent, that is, generating an agricultural surplus to support laborers and to provide materials for such a gargantuan construction project.

While archaeological research and restoration of pre-Columbian sites in the region have been minimal, the remains of countless terraces and check dams (*lama bordos*, or *jollas*) on adjacent slopes and arroyo bottoms, now mostly unused and

8. This is not a definitive finding. We have repeatedly experienced respondents deliberately understating the number of native speakers and the frequency of use of the Indigenous language in towns identified historically with such Indigenous groups.

9. Document transcribed in Paredes Colín ([1910] 1953, 71). Early sources did not distinguish between Chocho and Popoloca, which are linguistically close, so the language heard in Coxcatlán may well have been Popoloca.

10. Castillo Tejero, ca. 2008, n.p.

11. Rincón Mautner 2015, 179.

12. Terraciano 2002, 475n.

13. Paredes Colín (1910) 1953, 31.

14. The convent fell into ruin in the nineteenth century. It has since been restored and overwhelms the rest of the town's structures.

FIGURE 5.1 Coixtlahuaca convent, constructed under the supervision of Dominicans. Begun in 1545, it required fifty years to build. The labor to construct this masterpiece was provided by local Chochos, who neglected their milpas to serve the priests.

riven with collapse and erosion, attest to the lengthy and productive agricultural history of the town and its environs. Recent government-sponsored re-creation of *lama bordos* and terraces has imitated the ancient technique and created hundreds of kilometers of water and soil-retaining structures. For the ancients, these structures permitted expanded agricultural production and water retention. Two millennia ago, the Mixteca Alta may have supported a greater population than Oaxaca. Coixtlahuaca alone may have covered a greater agricultural area than that of the Valley of Oaxaca.[15] Now, however, the surrounding hills to the north and east appear from a distance to be nearly pure caliche, or tepetate, a hardened volcanic soil, and devoid of topsoil. The terraces, clearly visible in nearly all directions, are a mix of pre-Columbian structures and modern attempts at erosion control. They represent an ancient response to the loss of soils from those hills, an attempt to retain soil, moisture, and crops in a land perilously subject to topsoil erosion (for further discussion of erosion of the Mixteca Alta, see chapter 7). To the immediate south

15. Pérez Rodríguez 2013, 80.

FIGURE 5.2 *Lama bordos* near Chicagua, Nochixtlán, Mixteca Alta, Oaxaca. The structures prevent erosion, trap sediments, and store water, producing ideal sites for milpas.

and west of Coixtlahuaca, the hills are highly erodible sediments and resemble vast though colorful badlands. The drainageway from the south is a deep, recently incised channel that suggests (but does not entail) rapid erosion in recent decades.

The appearance that Coixtlahuaca and its *agencias* present, one of depopulated and economically depressed towns and villages, belies the once robust economy and vigorous artisan center as described in Aztec *relaciones*. As if to underscore the environmental deterioration of the region, Mexican census records show that the population of the Coixtlahuaca district (comprising thirteen counties) declined from more than eighteen thousand in 1950 to just over nine thousand in 2020.[16] Crops were once sufficient to support a large and well-educated population, one capable of supplying the host of skilled masons and sculptors needed for constructing the massive convent. Coixtlahuaca now appears to have more abandoned buildings than those occupied.

Although linguists make a clear distinction between Chocho and Popolocan languages, the two groups are often difficult to distinguish historically. Suffice it to say

16. Census information from Instituto Nacional de Estadística, Geografía e Informática (INEGI), accessed on August 16, 2021, http://www.inegi.org.mx.

FIGURE 5.3 Satellite photo of terraces and *lama bordos* near Nochixtlán. Google Earth.

that in general, Chochos occupied highlands in the Mixteca Alta directly adjacent to those of Popolocas, who occupied lands below, in the Tehuacán and Zapotitlán Valleys, and in some regions to the north and west.

Cuicatecs. Most of the towns on the western slopes of the Sierra Juárez above (to the east of) La Cañada and Cañada Chica (Río de las Vueltas) are home to Cuicatec speakers. Historically, Cuicatecs also lived in Cuicatlán and Quiotepec (both Aztec names). Both towns are now almost entirely Spanish speaking, except on market days in Cuicatlán, when one frequently hears the Cuicatec and Mixtec languages spoken. Cuicatec potters and peasants arrive there from their villages to the east in the Sierra Juárez to sell their wares. Mixtec basket and hat weavers arrive from the western slopes, the Sierra Mixteca. Both traditions probably predate the Conquest. Quiotepec, 20 kilometers north, also has a multiethnic deep history, which is not surprising, given its strategic location. It has historic sequences of occupation by Cuicatecs, Zapotecs, Cuicatecs again, Nahuas (probably Aztecs), and Mazatecs. Other occupiers may well have enjoyed the location as well during its nearly three thousand years of formal settlement.

Cuicatec speakers (they numbered over ten thousand in 2010) are now mostly concentrated in the mountainside villages of San Juan Tepeuxila and Teutila (which lies beyond the limits of La Cañada), as well as in four mountainside towns named Pápalo

FIGURE 5.4 Concepción Pápalo, a Cuicatec town with a Náhuatl name. It is situated on the western slopes of the Sierra Juárez well above Cuicatlán.

(which means "butterfly" in Náhuatl), the largest of which is Concepción Pápalo. The towns and villages are scattered over the steep slopes overlooking La Cañada and the Río Grande. In addition, at least two Cuicatec villages are located on the eastern slopes of the Sierra Juárez. The town of Cuicatlán, home to about ten thousand inhabitants, is located on gentle slopes overlooking the Río Grande. It is now a largely mestizo city, but it has regional importance, and its weekly market is the largest for many kilometers around. Cuicatecs and Mixtecs journey to the market each week, to shop and to bring wares to sell or barter from their villages in the nearby mountains. In line with the Náhuatl name (Place of Songs), music is everywhere in town.

Cuicatec villages are impoverished. Some residents, primarily women, earn a pittance by producing ceramic platters, which they market in Tehuacán. Mexico's national census bureau, INEGI, has pronounced the Cuicatec municipio of San Juan Tepeuxila the poorest in all of Mexico, with more than 97 percent of its inhabitants living below the official Mexican poverty level.[17]

For much of the human history of La Cañada, Cuicatecs made the valley bottom their home. Outside forces repeatedly drove them to relocate to the mountainsides,

17. Instituto Nacional de Estadística, Geografía e Informática 2010.

FIGURE 5.5 Corn and sugarcane fields and mango trees in Cuicatlán, with cliffs of the Oaxaca Fault. Former Cuicatec subsistence farmers and their milpas were driven from these fertile bottomlands to the mountainsides so that the new owners could plant commercial crops.

most recently Spanish settlers, who commandeered the best valley bottomlands for their own uses and relegated the Cuicatecs to locations above the valley to the east. Cuicatec *cacicazgos* have a long and complex history, linking contemporary towns with pre-Columbian settlements.

Ixcatecs. Santa María Ixcatlán is home of the few remaining Ixcatec speakers. The language is closely related to Chocho and Popoloca, but only distantly related to Mixtec. Rincón Mautner believes Ixcatecs split from Popolocas about 700 CE and then migrated to their present location.[18]

The town is situated on an isolated dry hillside of volcanic soils adjacent to limestone and calcareous soils, all overlooking a drainage high above La Cañada. Coixtlahuaca of the Chochos and Cuicatlán of the Cuicatecs are roughly equidistant from Ixcatlán, but the Mixtec village of San Pedro Nodón appears to be strategically located between Ixcatlán and La Cañada.[19] This positioning suggests that Mixtecs

18. Rincón Mautner 2015, 178.

19. San Pedro Nodón (see figure 4.19) is now a Mixtec village, according to its inhabitants. The French explorer León Diguet, writing in the early twentieth century, pronounced it "exclusively" Popolocan. See Jäecklein 1979, 201.

FIGURE 5.6 Santa María Ixcatlán. Once the prosperous center of the Ixcatec state, it is the only Ixcatec town remaining. In 2010 only ten speakers of the original language remained.

and Chochos surrounded the Ixcatecs. Whether hostilities erupted among the three groups is difficult to determine, but the fact that the Ixcatec language evolved in this single city/region suggests that its isolation continued for centuries. Ixcatecs formerly occupied several nearby villages, but in the last centuries, the lowering of the water table combined with the difficulty of practicing subsistence agriculture has forced the congregation of the remaining ethnic members into Santa María itself, a town with many empty buildings.[20]

The Ixcatlán region, like Coixtlahuaca to the southwest, has experienced extreme soil erosion and degradation in recent centuries, and what was once a thriving agricultural realm is now a vast, eroded *palmar*, vegetation consisting overwhelmingly of diminutive palm trees of the genus *Brahea* and other chaparral plants, plus a sprinkling of glades of low-growing oaks. Milpas are rather few, though the adjacent hillsides exhibit clear signs of ancient terracing. The principal industry in the town is the weaving of simple hats and baskets from palm for resale in cities, especially Mexico City. A few men still produce mezcal, but they complain that the lack of

20. Byers 1967, 62.

FIGURE 5.7 Hat weavers, Ixcatlán. The hats sell for a pittance in Tehuacán and Mexico City.

FIGURE 5.8 Mezcal producer, Ixcatlán. Piñas, the starchy cores of magueys (agaves), will be roasted, chopped into smaller pieces, soaked, and fermented. The liquid will be distilled into mezcal.

FIGURE 5.9 Hillside in Mazatec lands, Sierra Mazateca. On lower slopes, Náhuatl speakers live in hamlets.

water limits production. Residents have reported to us that fewer than ten Ixcatec speakers remained in 2012, the youngest of whom was over forty years old.

Mazatecs. Mazatecs do not appear to occupy any settlements in La Cañada or the Valley of Tehuacán. Centuries ago, however, they apparently controlled Quiotepec and Tecomavaca, as well as perhaps Teotitlán de Flores Magón.[21] They may have also inhabited Coxcatlán around the time of the Conquest. Their principal territory is the moist cloud forests (or lands formerly forested) and eastern lowlands of the Sierra Juárez north of Chinantec country and the Sierra Mazateca east of Teotitlán, extending into the tropical moist lowlands of northern Oaxaca and southwestern Veracruz. Mazatecs regularly appear in the market of Teotitlán. Driving northeast, up the steeply rising, sinuous Mexico Route 182, one passes through several hamlets where Náhuatl is spoken, soon arriving at Mazatec settlements. (Linguists recognize eight different languages in the Mazatec group, with around two hundred thousand speakers as of 2013.[22]) It is easy to imagine that Teotitlán (a Náhuatl name) and environs were once part of Mazatec territory.

21. Hunt 1972, 213.

22. Lewis, Simons, and Fanning 2013.

FIGURE 5.10 Cuicatlán and the Oaxaca Fault viewed from the west on a roadway near San Pedro Nodón.

Mazatecs' homelands may have once been immediately adjacent to those of Cuicatecs, but that joint boundary appears to have been penetrated by Chinantecs who arrived from the east many centuries ago and drove a partial wedge between them. Huautla de Jiménez is a Mazatec town. Near Huautla, the Mazatecs' use of *Psilocybe* mushrooms, and more recently, *Salvia divinorum*, has attracted many foreigners in search of alternate viewing of the cosmos. From a viewpoint nearby, one can see mountainside Chinantecan villages to the southeast and Cuicatec villages to the southwest. As of this writing, any Mazatec archaeological sites in the La Cañada region remain to be described. Elsewhere, caves in limestone formations are home to extensive archaeological records, forty-eight of them on the prominent landmark Cerro Rabón.[23] For many Mazatecs, that massive limestone mountain that rises abruptly from the coastal lowlands is emblematic of their ancestral lands.

Some linguists and historians have proposed that Mazatecs were among the most probable founders of Teotihuacan.[24] This view is not widely shared.

23. Hapka and Rouvinez 1997, 22–25. The Sierra Mazateca is home to some of the world's largest cavern systems.

24. Paddock 1997, 106.

Mixtecs. Of all the linguistic groups now represented in the reaches of La Cañada and the Valley of Tehuacán, the Mixtecs (or Ñuudzahui—"people of the rain"—as they refer to themselves), figure most prominently in pre-Columbian history.[25] This is especially true of some sites located in the Mixteca Alta, which forms the western boundary of La Cañada and the southwestern drainages of the Valley of Tehuacán. Only a small percentage of Mixtec territory lies within La Cañada or Valley of Tehuacán drainage.

Linguists recognize as many as fifty-two languages in the Mixtec group, a number that symbolizes the challenges facing ethnohistorians seeking to reconstruct Mixtec history.[26] Indeed, compiling comprehensive descriptions of the region and its history has occupied the lifetimes of many archaeologists. The daunting diversity of languages becomes understandable with a satellite view of the region, which reveals a mosaic of remote small valleys surrounded by (often) forested ridges and peaks. The peculiar ecological conditions of these valleys—generally abundant summer rainfall, cool climates with winter frosts, and numerous pockets of fertile soil—meant that many of them could easily become self-contained and socially isolated from one another while supporting a large population. Indeed, prior to the Classic period, the Mixteca Alta supported the largest population in all highland Mesoamerica, more even than the Basin of Mexico.[27] The populations also swelled during the Postclassic.

The modest valleys of the Mixteca Alta gave rise to the Mixtecs' basic pre-Hispanic social settlements. They were the *ñuu*—"potentially autonomous states"—groupings of smaller barrios into a larger community. Each of the many valleys, large and small, constituted a *ñuu*. "The ñuu was larger than a community; . . . it was a multi-community grouping. The ñuu was an integral autonomous unit or an integral component of a larger polity. It was not dissolved in a more centralized solution. It retained its integrity even when drawn into alliance or core-periphery relationships."[28] The *ñuu* might form or be drawn into an alliance with other, more powerful centers, but once the power was relaxed, the *ñuu* would retreat to its former autonomy. This fundamental unit remained (and to an extent remains) the bedrock of social organization of the Mixteca Alta and helps explain the relative absence of emergent imperial powers among the Mixtecs. It also makes more understandable the failure of Monte Albán to conquer the Mixteca Alta as it conquered La Cañada. The individual *ñuu* might be easy to conquer, but there were many, and ruling over all of them would challenge all but the most powerful states, and none in the history

25. Mixtec is the Nahua word given to the Ñuudzahui by the Mexica, meaning literally, people of Mixtlán, people of the clouds. Burgoa (1674) 1934.

26. Lewis, Simons, and Fanning 2013.

27. C. A. Smith 2002, 121; Spores and Balkansky 2013, 89.

28. Kowalewski et al. 2009, 337.

of pre-Columbian Mesoamerica was up to the task. Aztecs defeated the larger *cacicazgos* but left life uninterrupted in much of the region. Despite the proliferation of small states, the artistic and architectural accomplishments of the Mixtecs of the Mixteca Alta are vast and impressive.

Some Mixtec polities in Oaxaca became regionally powerful toward the end of the Classic period. As the might of the Zapotec state waned following the decline of Monte Albán, that of the Mixtecs expanded, and by the seventh century CE, Mixtecs appear to have developed brisk trade and political influence northward at least far as Cholula, and they began to expand to the south as well. Paddock believes that the Mixtec *cacicazgo* of Tilantongo became involved in the conquest of Cholula in 720 CE and three centuries later was able to unite all or most of the *cacicazgos* of the Mixteca Alta and Tututepec on the Pacific coast under one rule.[29] Burgeoning trade routes connected the Mixteca Alta with the Valley of Zapotitlán and north into the Basin of Mexico, but they probably bypassed the Valley of Cuicatlán entirely and much of the Valley of Tehuacán as well. Centuries later, a few decades prior to the Spanish Conquest, the Mixtec kings became military allies of the Zapotecs in the attempt to ward off Aztec imperialism.[30]

The long-term subordination of Mixtecs to Zapotecs, if it indeed occurred, is perhaps attributable to the absence of anything approaching a major urban center in the Mixteca Alta. Instead, in addition to the *ñuu*, the Mixtecs appear to have developed a series of religious centers that held sway over a self-sufficient area of settlement without any special concentration of military power. Yucuita, for example, supported a large population and was surrounded by a sizable wall but does not seem to have exerted an urban influence over the surrounding region, apart from the protection provided by its perimeter walls.[31] The same is true of Tilantongo in the seventh to eleventh centuries CE, despite its military forays to the north and south.[32] Though several culturally and economically successful *cacicazgos* emerged in the Mixteca Alta over the centuries, all appear to have been decentralized concentrations of residents—constantly changing assemblages of *ñuu*—rather than densely populated urban centers.[33]

How do we explain the limited urban development in the Mixteca Alta and the Mixtecs' peculiar style of community organization? To some extent, we can look

29. Jiménez Moreno 1966, 63. The author cites as his source the annals known as the *Historia Tolteca-Chichimeca*, also referred to as *Cuauhtinchan*. Paddock 1966, 202, 205; see also Spores and Balkansky 2013, 105–6.

30. Marcus and Flannery 2000, 400–401.

31. Balkansky 1998, 467.

32. Paddock 1966, 202–5.

33. Spores 1967, 92–102.

to the comparatively low yields of milpas in the hilly region and the cool climate of the Mixteca Alta with its shallow, easily erodible (though fertile) soils. The flatland valleys of the Zapotecs in Oaxaca's central valleys, with deep, stable soils and warmer climate, were more productive and easier to farm. While Mixtec productivity increased vastly with the construction of thousands of terraces and *lama bordos*, or *coo-yuu*, as in Coixtlahuaca, these were built and maintained at great labor cost. Despite the vast area amenable to planting in the Mixteca Alta and the numerous pockets of soil with enviable fertility, frosts are not unusual, and cool temperatures pose a limiting factor on yields. Only one crop of corn is possible in much of the Mixteca, while double cropping is possible in La Cañada and in some parts of the Oaxaca valleys.

So the absence of urban centers in the Mixteca Alta is attributable to several factors. First, comparatively low crop yields made difficult any surplus to support an urban population.[34] Second, the reported high population of the Mixteca (which came to rely on the *lama bordos* for food),\ was dispersed and fragmented into small townships, *ñuu*, and *cacicazgos* that appear to have cherished autonomy and resisted concentration in any urban center. Third, the ecological cost of monumental architecture, the extraction of trees to roast limestone for making plaster and mortar, and the diversion of labor from farming to building construction further limited agricultural surpluses. Furthermore, rapid population decline due to disease following the Spanish Conquest cast a negative effect over any urge toward urban concentration from Mixteca Alta *cacicazgos*. Centuries before the arrival of Spaniards, however, construction of monumental buildings appears to have come to a halt, even though population did not decline substantially until much later, after the conquest of the Mixteca Alta by Spanish forces.

Still, the Mixteca Alta was for two millennia an active site of alternately burgeoning and waning *cacicazgos*. Though typically small in area, these statelets exhibited brilliant cultural and architectural achievements. As travelers passed through La Cañada after roughly 500 BCE, they would have encountered little in the way of temples, pyramids, raised platforms, or palaces. The lands above the river 20 or so kilometers west, however, teemed with monumental architectural activity and supported a large contingent of talented stone masons, painters, and ceramicists. Mixtecs were closely related to Popolocas to the north and exerted strong influence on Popolocan culture in the Valley of Zapotitlán, perhaps as late as Postclassic times. Several trade routes connected the Mixteca Alta with trade centers to the north, but

34. And yet agricultural surpluses were enough to support the huge labor force necessary to construct the massive convent structures at Coixtlahuaca, Teposcolula, and Yanhuitlán under the Dominican priests in the sixteenth century.

for other peoples, the principal route between the Basin of Mexico and Monte Albán bypassed the Mixteca in favor of La Cañada.

The town of Apoala has been mentioned in codices as a cradle of Mixtec civilization.[35] It was certainly a *ñuu*, but no intensive archaeological studies have been published to bolster its claim to the origin of the grand nobility. The Codex Vindobonensis mentions that One Deer created both the world and Mixtec nobility, who emerged from the earth via a mythical tree growing in Apoala, recognized as the ancestor of all Mixtecs (see figure 4.18).

One can believe the essence of the myth when first viewing the valley from an overlook, for it is one of transcendent beauty and pastoral tranquility (see figures 4.16 and 4.17). The valley has fertile and easily irrigated farmland, but acreage is limited to the valley bottom and would support a population of only a few hundred. Mixtec is spoken in Apoala and in numerous *sujetos* (hamlets) in the surrounding area. (Mixtec has so many languages and dialects that inhabitants of one village may speak slightly or considerably differently from another only a few kilometers distant.) Viewing the valley from above, it is not difficult to envision it as a small *ñuu*, an enduring polity, autonomous, but willing at times to forge alliance for protection or economic advantage.[36] And such is the nature of the Mixteca Alta, a region of numerous pockets of (at least once) rich land in a matrix of steep, shallow-soiled hills that must be terraced to farm, ideal for giving rise to small polities isolated by ridges and foreign languages. The canyons that descend from the slopes are also havens of biological diversity, leading to the creation of new, closely related species by geographic isolation.

The Mixteca Alta, the geographic region of Oaxaca of which Apoala, Coixtlahuaca, and Nochixtlán are all part, is widely known for its chief export, namely Mixtecs driven out by the impoverishment of the land, changes in government agricultural policy, and population pressure. Large colonies of expatriate Mixtecs can be found in Mexico City, Baja California, and Los Angeles, where they maintain their traditions and their languages and nourish close contacts with the villages of their origins in Oaxaca.[37]

The causes of the exodus of Mixtecs are readily apparent in the landscape of the Mixteca Alta. Apart from a few locations, such as the lush small valley of

35. Jiménez Moreno 1966, 61; Spores 1967, 100.

36. For Spaniards' colonial purposes, they imposed on the Mixteca Alta the *cabecera-sujeto*-barrio system of town and city administration based on Spanish political organization: *cabecera*, administrative center; *sujeto*, administrative dependency; and barrio, neighborhood with certain governmental status. This system not only obscured the nature of sociopolitical relationships among *yuhuitayu*, *ñuu*, or *siqui* (similar to a barrio) but also did not accurately represent actual colonial settlement patterns. See Terraciano 1994, 119–24.

37. For an analysis of the exodus of Mixtecs from the Mixteca Alta, see Edinger 1996.

FIGURE 5.11 Downstream from Apoala showing canyon habitats. The Sierra Mazateca is the large misty range in the distance.

FIGURE 5.12 Mixtecs selling and buying palm *cogollos* (young fronds) for weaving, Nochixtlán.

Apoala, much of the remainder of the adjacent country is deforested, grotesquely eroded, and currently incapable of supporting the large numbers to which it was once home. This impoverishment of the countryside can in turn be traced, at least in part, to an economic system that for centuries undermined self-sufficiency.[38] This is especially true of the eastern portion in the vicinity of the small cities of Coixtlahuaca, Nochixtlán, and Yanhuitlán. (The latter two lie just outside the drainage of La Cañada.) Early Spaniards describe the region as prosperous and heavily populated, so much so that it became the object of intense evangelization by Dominican priests in the early sixteenth century. Under the monks' supervision, Mixtecs and Chochos constructed three enormous convents with chapels that remain historical monuments and a tribute to their artistic talents, in Coixtlahuaca, Teposcolula, and Yanhuitlán. Agriculture and industry in the region were

38. The French writer François-René de Chateaubriand supposedly wrote, "Forests precede civilizations and deserts follow them." It is an apt quotation but wrongly attributed to Chateaubriand. See Le Bot 2012. As in many other parts of the Americas, the fate of the land by successive waves of conquest and exploitation was sealed with the arrival of livestock, mainly cattle, goats, and sheep. Mexican researchers estimate that 80 percent of the soils of the Mixteca Alta are severely eroded (Guerrero-Arenas, Jiménez Hidalgo, and Romero 2010, 66). While the most glaring episodes of erosion appear to have taken place following the Conquest, the widespread and early use of terracing in the Mixteca Alta indicates that its erosive soils were showing the effects of erosion on the population three thousand years ago. Pérez Rodríguez and Anderson 2013, 335–39.

FIGURE 5.13 Market in Nochixtlán, Mixteca Alta.

sufficiently robust to support large populations in numerous *ñuus*, especially in the Classic period. when the population probably exceeded one hundred thousand. At that time Mixtecs were renowned for their exquisite artisan productions and their architectural sophistication, accomplishments that the Spanish priests noted and used to their considerable advantage. Dominicans wrote that Indians were inferior to Europeans in mentality and intellectual capacity, but the clerics seemed quite willing to incorporate the Mixtecs' skills into some of Mexico's finest post-Conquest architecture.

Nochixtlán and other important towns of the Mixteca Alta are located on higher, more rolling country southwest of Apoala.[39] For traders or soldiers ascending westward from La Cañada, they would be the first heavily populated areas to be encountered.

Coixtlahuaca, to the northwest, as we have seen, was probably a mixed Chocho-Mixtec statelet. Aztec narratives describe the conquest and destruction of several

39. Nochixtlán, one of today's most important Mixtec centers, translates from the Náhuatl as the bland name "Place of the cactus." Coixtlahuaca, Tamazulapan, Teposcolula, and Tlaxiaco are all Náhuatl names. Broadly speaking, Nahua refers to any Náhuatl speaker. These included myriad groups across Mesoamerica: as far north as Durango, and southward to Guatemala, Salvador, Nicaragua, Costa Rica, and Panama (called there Pipil, Nicarao, and Guatuso). A large geographic hiatus in the distribution of Náhuatl is evident in Oaxaca and Chiapas.

Mixtec cities, and though the Aztec names remain, other influences of their occupation have vanished.

The role of the Mixteca Alta in the deep history of the Cuicatlán and Tehuacán Valleys is difficult to assess. The highlands are extraordinarily rich in archaeological sites and traditions, but their influences have been diffuse and complex. Their role in the human history of La Cañada, only a few kilometers distant from sophisticated ceremonial structures and cities of stratified societies, must be profound, yet clear evidence of direct involvement in the life of La Cañada is lacking.

The reason for this confusing situation may lie in the routes of communication between the Mixtecs and their linguistic relatives to the north, the Chochos and the Popolocas of the Valley of Zapotitlán. Castellón argues that the principal communications between Zapotitlán of the Popolocas and the Mixteca Alta were two routes, one following Mexico 125 southwest to Huajuapan de León, then southeast into the Mixteca Alta. The other route, shorter but more mountainous, passed through Los Reyes Metzontla and followed canyons now crossed by Highway 135D into the Mixteca Alta, arriving at Coixtlahuaca.[40] Either of these routes would be shorter than descending the Río Zapotitlán to its juncture with the Río Salado, descending the Río Salado to its junction with the Río Grande, then following La Cañada and Cañada Chica into the central valleys or up a steep side drainage to Coixtlahuaca. Furthermore, the routes through the Mixteca Alta (thus bypassing the valleys) would take travelers into areas linguistically familiar to Popolocas (and in reverse direction, to Chochos and Mixtecs), whereas the Tehuacán-La Cañada route passed through areas occupied first by Nahua speakers, then Mazatecs, Cuicatecs, Chinantecs, and finally Zapotecs. Mixtecs naturally gravitated to the north, not to the east.

Nahuas. Náhuatl, the language of the Nahuas, including the Aztecs, has by far the largest number of speakers of any Indigenous language in Mexico. Linguist Jorge Suárez believed that Náhuatl was spoken in central Mexico by 700 CE, but later linguists place the date perhaps a century earlier.[41] Classical Náhuatl, the language of the Mexica, was used as a lingua franca throughout Mexico and well into Mesoamerica from 700 CE until the Spanish Conquest in the sixteenth century. The modern dialects of Náhuatl spoken in the Valley of Mexico are closest to classical Náhuatl. At the time of the arrival of Spaniards, Náhuatl was widely spoken throughout Mesoamerica, a symbol of Aztec imperialism.

In many of the towns and small cities of the valley, both north and south of Tehuacán, Náhuatl is still spoken in public places. (Southeast of Tehuacán, the valley towns or small cities of Ajalpan, Altepexi, San José Miahuatlán, Santa María

40. Castellón 2009, 62–63.

41. Suárez 1983, 149.

FIGURE 5.14 Banner advertising study in Náhuatl theology, Altepexi, Valley of Tehuacán. Note the Aztec noble at the left of the banner and the Náhuatl glyph for Altepexi on the right. The banner announces another attempt by the Catholic Church to absorb Náhuatl beliefs into Catholicism.

Coápam, San Gabriel Chilac, and San Sebastián Zinacatepec are home to numerous speakers, and Náhuatl is commonly spoken in their markets.) The isolated town of San José Axuxco is also home to many Náhuatl speakers, though most of them appear to be adults. The village of San Pedro Tetitlán, some 20 kilometers south of Tehuacán, remains strongly Nahua. Residents there claim that all natives, including children, speak Náhuatl.

The mountains east of the Valley of Tehuacán—primarily the Sierra Zongolica—are home to numerous Nahua villages on the eastern and western slopes.[42] If we plot these on a map along with Popolocan locations, it appears that in pre-Conquest times, Popolocas constituted an impediment to southward and westward expansion by Aztecs and their Náhuatl-speaking predecessors. The Nahuas then drove the Popolocas off to the west and settled the Tehuacán Valley with colonists, perhaps populating the Sierra Zongolica at the same time. The specifics of that conquest are elusive, and the date of arrival of the Nahua population remains undetermined. Most authorities date the arrival of the Nahuas in central Mexico around 700–750 CE,

42. See, for example, Rodríguez 2003.

FIGURE 5.15 Nahua woman holding *tenate*, a basket for carrying tortillas, San Pedro Tetitlán, Valley of Tehuacán.

FIGURE 5.16 Nahua women preparing *sopes* of blue corn tortillas, Zinacatepec market, Valley of Tehuacán.

which would coincide with the decline in production of Thin Orange pottery at Tepexi de Rodríguez and with the late rapid decline of the influence of Teotihuacan. This date also corresponds with the Olmeca-Xicalanca conquest of Cholula, which, according to tradition, resulted in the expulsion of Nahuas, who became known as Pipiles and migrated to points south and east.[43]

The abundance of Náhuatl place names plus the location of towns where Náhuatl is still spoken provide a clear demonstration of the scope of Nahua penetration into Otomanguean territory. Uto-Aztecan languages are still spoken in a band that extends from Nicaragua to Idaho, but communities that speak the Náhuatl language (or any Uto-Aztecan tongue) appear to be absent from the central valleys and the Mixteca Alta of Oaxaca. Nahuas arrived there, conquered, extracted resources, and departed or were absorbed by the powerful, already ancient native cultures.

Popolocas. While Zapotecs, Cuicatecs, and Mazatecs were major players in the pre-Columbian Valley of Cuicatlán, Popolocas represent the most important identifiable pre-Aztec inhabitants of the Valley of Tehuacán. Investigators now agree that Popolocas were ancient residents of the valley and were responsible for most of its archaeological history, yet Popolocan culture was marginalized or ignored until the publications of Mexican historian Nicolás León in 1905.[44] Edward Sisson, who studied the archaeology of southern portions of the Valley of Tehuacán extensively in the 1970s, believed that the entire valley, including Coxcatlán, was once under Popolocan control.[45] More contemporary scholars believe lands beyond the valley to the west and north were under Popolocan control as well, and southern limits of Popolocan power may have reached beyond Coixtlahuaca to include Nochixtlán, Oaxaca, now considered by many to be the center of contemporary Mixtec culture.[46] The greater Popolocan area of influence would include Huajuapan de León, Oaxaca, in the Mixteca Baja; Acatlán, Puebla, and Tepexi, Puebla, all situated well beyond the limits of the Valley of Tehuacán.

The Valley of Zapotitlán, a side valley southwest of the city of Tehuacán, has for centuries been strongly Popolocan. The Popoloca language has close affinities to Chocho and the region of Coixtlahuaca in Oaxaca. It is well represented in the Zapotitlán Valley in the town of Los Reyes Metzontla and in several towns northwest of Tehuacán beyond the valley's limits. The town of Atzingo, in the southern valley of Tehuacán at the point where the Arroyo Zapotitlán joins the Valley of Tehuacán, is also strongly Popolocan. Today the women of Popolocan Atzingo embroider huipiles on a piecemeal basis for Nahua proprietors of wholesale and retail

43. See, for example, Cepeda 1997, 27.

44. Gámez 2003, 9, 69ff; León 1905.

45. Sisson 1997, 130.

46. Gámez 2003, illustrations, esp. fig. 17; Castillo Tejero 2000, 299.

textile shops in Náhuatl-speaking San Gabriel Chilac, a few kilometers away. Their role as economic subordinates perhaps indicates a vestige of Popolocan subservience to Aztec imperialism. North of Tehuacán, the town of San Luis Temalacayuca retains a large Popoloca-speaking population, though of a different dialect from that spoken in Atzingo. Historically the city of Tecamachalco, positioned at the very northern end of the Valley of Tehuacán was a Popolocan stronghold.[47] Residents of Zapotitlán Salinas, the dominant town in the Valley of Zapotitlán, identify their heritage as Popolocan, but few speakers, if any, reside in the town. Archaeologists believe that the Popolocan town of Tepexi de Rodríguez, northwest of the Valley of Tehuacán was the home of the celebrated Thin Orange ceramics, which became widely circulated throughout Mesoamerica between 300 and 750 CE.[48] The settlement came under the influence of Teotihuacan, which marketed the ceramics. This production center ties Popolocan country firmly into the prehistory of Mesoamerica, though the ethnicity of the inhabitants during the Classic period has not been determined. An independent and much later Popolocan stronghold, Tepexi el Viejo, 6 kilometers from the town, appears to have remained independent until Aztecs conquered it in 1502.[49]

Some archaeologists believe that Popolocas formed an influential presence in Teotihuacan, which would date their cultural presence back into the Early Classic period.[50] Paddock suggests that Popolocas from barrios in Teotihuacan may have returned to their homelands after the fall of the great city, bringing Teotihuacan influences with them.[51]

Other traditions suggest that Tehuacán was founded in an area probably settled by Popolocas early in the twelfth century CE. Alejandra Gámez has interpreted codices and oral histories to establish that Popolocas arrived in the vicinity of Tecamachalco at the northern end of the Valley of Tehuacán in 1179 CE from their early origins in Coixtlahuaca, now known to have been a Chocho enclave.[52] (This does not preclude their ancestors from having spent time in Teotihuacan.) Her interpretation suggests that Popolocas penetrated as far north as Cuauhtinchan, in the vicinity of the city of Puebla, and intermarried with Nahuas, probably descendants of Tula. They proceeded to found an independent polity at Tepexi el Viejo, in the same region where, centuries earlier, their ancestors had produced the Thin Orange pottery in an adjacent settlement. Much later they formed the government of Tecamachalco

47. Gámez 2003, 181–97.

48. Rattray 1990, 181–95.

49. Jäecklein 1979, 200.

50. Jiménez Moreno 1966, 43.

51. Paddock 1966, 43.

52. Gámez 2008, 72.

and the allied polity of Quechaca (Quecholac) to the east and strengthened into a regional power.[53]

Historian Joaquín Paredes Colín cites Franciscan archives in pronouncing the natives of Tehuacán to have been Popolocas and to have been especially receptive to evangelizing by the Franciscans. For several centuries, then, Popolocas appear to have occupied the Valley of Tehuacán as far south as Coxcatlán, which would make them immediate neighbors of Cuicatecs and/or Mazatecs and quite possibly residents of the city of Coixtlahuaca in the Mixteca Alta.[54]

Given the broad and persistent distribution of Náhuatl speakers in the Valley of Tehuacán, however, Popolocas of the Late Classic and Postclassic may have governed only with the consent of Nahua powers to the north, especially Cholula. This power-sharing arrangement would have been disrupted as Aztecs, the late-arriving Nahuas, became the universal power of Mesoamerica by the beginning of the fifteenth century CE, and it would have ended with the conquest of Popolocas by the army of Moctezuma I in the mid-fifteenth century. The documents Paredes Colín cites suggest that at the time of the Spanish Conquest, Coxcatlán was a Nahua town, not Popolocan, and had been for some time.[55] A broad intrusion of Náhuatl speakers into the Valley of Tehuacán remains in place, extending south to Teotitlán de Flores Magón, Oaxaca, and terminates abruptly there, only to be resumed far to the southeast in El Salvador.

The extensive archaeological site of La Mesa (Tehuacán Viejo) is located atop a mesa on the eastern side of Tehuacán, near the Nahua town of San Diego Chalma (see figure 6.21). Popoloca-speaking towns lie on the western side of the Valley of Tehuacán, so if, as archaeologists believe, the site is of Popolocan origin (or of a mix of Otomanguean-speaking groups including Popolocas), then Aztecs or Nahua predecessors of Aztecs forced the Popolocan populations to the west at some point *after* La Mesa was constructed. The conquerors then confined most of the vanquished peoples of the Valley of Tehuacán to the western part of the valley and the Valley of Zapotitlán, perhaps encouraging them, we suspect, to confine their activities to producing salt and food for export.

While it is tempting to view the profusion of Náhuatl-speaking towns in the Valley of Tehuacán as a heritage of Aztec imperialism, care must be taken to consider the complex origins of the Náhuatl language. Náhuatl speakers abounded in the

53. Popolocan occupation of the Valley of Tehuacán may date much further back in history: Ceramics excavated near Purrón Cave that date from more than two millennia BCE are remarkably like those produced today at Los Reyes Metzontla in the Valley of Zapotitlán. The village still produces attractive but utilitarian ceramics 2,300 years later. Gámez 2003, 73.

54. Paredes Colín (1910) 1953, 30–32, 71; Castillo Tejero ca. 2008.

55. Paredes Colín (1910) 1953, 69–70.

FIGURE 5.17 Popolocan potters producing comales (tortilla griddles), Los Reyes Metzontla, Puebla.

Basin of Mexico long before the rise of Tenochtitlán, and Náhuatl is still the dominant language in much of rural Puebla to the north. As we have seen, spoken Náhuatl preceded the Aztecs by as much as eight hundred years. In other words, various peoples spoke Náhuatl long before the rise of the Aztecs.[56] Castillo Tejero suggests that Nahua speakers invaded Popolocan country at least two hundred years before the arrival of Aztecs.[57] These, not Aztecs, may be the ancestors of contemporary Nahuas in the Valley of Tehuacán.

Popolocas, like most other Indigenous Mexicans, are marginalized from wealthier Mexican society, hardly different from the way León found them, impoverished and oppressed in 1905. Centuries earlier Spaniards found them to be a sophisticated people, noted for their ceramics and, remarkably, their soldiers' ability to leash four arrows at the same time.[58] Their monumental architecture itself reveals their intel-

56. Jorge Suárez (1983) dates Náhuatl presence in central Mexico to at least 700 CE. Aztecs, more properly Mexica, were the last of the seven mythical Náhuatl-speaking tribes (traditionally Xochimilca, Chalca, Tepaneca, Acolhua, Tlahuica, Tlaxcalteca, and Mexica), arriving in the Mexico Basin from traditional locations called Aztlán and Chicomóztoc. They arrived at a basin where Náhuatl was already a widespread language. Navarrete Linares (2011), in a deep study of the Mexica pilgrimage, discusses historical documentation of the many Nahua polities (*altépetl*) in the Valley of Mexico.

57. Castillo Tejero 2004, 174; Plunket and Uruñuela 2005, 104ff.

58. León 1905, 7.

lectual and artistic achievements. Today, most men work as subsistence farmers, pastoralists, or day laborers, supplementing their milpa products with palm weaving and pottery making, industries in which they are assisted or supplanted by their wives or other family members. Some own small businesses. Ceramicists of Los Reyes Metzontla appear to produce pottery using a technology largely unchanged in two millennia, much as León described it. Women are the primary potters. The men in their families provide them with clay and the materials for producing desirable colors. The latter requires a sophisticated knowledge of mineral and vegetable sources of coloring.[59] While the pottery is largely utilitarian, it bears an unmistakable Metzontla style and is curiously appealing.

Pre-Columbian Boundaries in the Valley of Tehuacán

A map illustrating the lands controlled by various Indigenous groups over the centuries in pre-Columbian times would reveal boundaries in an amoeba-like state of flux. The arrival of Nahuas in roughly 700 CE had an immediate and dramatic effect on ethnic land tenure in the region. Still, the boundary between pre-Aztecan Nahuas and other Otomangues besides Popolocas seems to have varied considerably according to the political and economic relations of bordering groups.[60] Aztecs probably introduced territorial stability unknown for millennia and fixed the boundaries in roughly their present location. At times the sparse archaeological record, the complex history of the area, the cycles of conquest and repossession of territories, and the fragmentary nature of written records and codices make it difficult to distinguish between internal regional hostilities among peer polities and wars with alien peoples. Were the disputes between Mixtecs and closely related peoples, perhaps Cuicatecs or Mazatecs, or were they battling Nahuas? Were the Mixtecs' defensive strategies directed at outsiders such as Nahuas or Zapotecs, or were they conceived as protecting them from neighboring Mixtecs or nearby Popolocas?

The answer seems to be that some defensive maneuvers, such as erecting perimeter walls and concentrating cabeceras in urbanized areas, took place without major or long-term effects on economic exchange; that is, no sign of conquest appears in the historical record until the arrival of Aztecs in the fifteenth century. Mixtecs may have constructed defensive works against possible Cuicatec or Zapotec aggression

59. Reynoso 1997, 122–26.

60. Navarrete Linares 2011, 30. "En general, las fuentes del valle de México coinciden en que los chichimecas hablaban otomí, o lenguas emparentadas con ese idioma." The Otomí language is related to many of the languages of the valleys and their environs.

while still conducting vigorous trade with the same potential enemies. Threats of conquest by a major foreign military alliance, however, appear to have affected both settlement continuity and commercial activity.[61] Whatever the picture that future archaeology provides us of the prehistory of the valleys, we can be certain that Aztecs and their Nahua predecessors successfully invaded lands occupied by Otomanguean speakers and left behind a landscape bearing the place names indicative of continuing conquest.

61. Plunket and Uruñuela 2005, 110.

CHAPTER 6

Archaeology

Since the inauguration of the Tehuacán Project in the 1960s, archaeologists have demonstrated increasing interest in the pre-Columbian history of the Valley of Tehuacán. While the goal of that project was to trace the agricultural history of the region, associated studies have brought to light ancient sites that otherwise might have remained unstudied. Fragments of ancient peoples have been unearthed and analyzed, and at least some excavation has been carried out since that time in several notable localities. These include (from south to north) Llano Perdido (La Coyotera) in La Cañada; Tecomavaca, Los Cues, and Teotitlán de Flores Magón, all in Oaxaca; and Coxcatlán, Calipan, and Tehuacán Viejo in the Valley of Tehuacán and Cerro Cuthá in the Valley of Zapotitlán, all in Puebla. Most of these sites have been ransacked and are overgrown with vegetation, as well as being difficult to discern. It is fair to say that in both valleys, every contemporary city or town is located on or near a pre-Columbian archaeological site, with the possible exception of the old railroad villages, such as El Parián and Valerio Trujano in La Cañada. By 1972 the Tehuacán Project had identified 456 archaeological sites in the Valley of Tehuacán, most of them from Classic and Postclassic times.[1] In 1971–72 Sisson and his associates added an additional sixty-three sites from Coxcatlán and environs alone.

Major archaeological sites have not survived well in La Cañada, probably because of the limited building area, expanded agriculture, and lack of a pre-Columbian population large enough to supply the skilled and large labor force for constructing major civic and ceremonial sites. It is also possible that Zapotec imperialism suppressed the development of a Cuicatec culture, preventing the coalescence of a social

1. Sisson 1973, 4–6. Sisson noted that the sites had been severely looted or destroyed by agricultural expansion. The situation today, more than forty years later, is even more grim.

structure of sufficient magnitude to develop monumental structures. Lest the reader consider the area to be archaeologically bereft, however, we note that within 20 kilometers or so to the west lie dozens of important Mixtec sites. Furthermore, Spencer and Redmond, in their initial survey of La Cañada, identified ninety-three sites.[2]

We focus on four archaeological sites. All four are contained within the Valley of Tehuacán, though one, Cerro Quiotepec, lies nearly within La Cañada and was closely aligned with Monte Albán in Oaxaca, whose rulers appear to have dictated its construction. For three of these sites, studies have been published but are not readily available to the public. In the case of Tehuacán Viejo (La Mesa), access to and documentation and descriptions of the site have only recently become accessible. For decades it was closed to visitors despite its scope and magnitude, which are such that failure to mention it would be a major sin against the valleys. It has since become one of Mexico's most visited archaeological spectacles.

We begin with the site of Cerro Cuthá in the Zapotitlán Valley because it is self-contained and accessible to the public. A hike of a mere three hours is sufficient to view the ruins and assess their position in the valleys, and local guides from Zapotitlán Salinas are well trained and eager to accommodate visitors. Of the other three sites, Quiotepec, though grand and superbly located, is even more remote and access more restricted, while Purrón is more difficult to locate and identify without a local guide. Tehuacán Viejo, the most important Popolocan ceremonial center, closes our archaeological reconnaissance. The siting of Tehuacán Viejo (La Mesa) is not haphazard. Located high at the northern end of the valley, it offers commanding views of the entire Tehuacán Valley, far toward the Oaxacan highlands, with a line of sight to Cerro Cuthá and the saltworks on the western edge of the valley.

Cuthá: An Ancient Popolocan Hilltop Town

Through settlement or conquest, the Aztecs and their predecessors transformed the Valley of Tehuacán into a southern bastion or perhaps a colony of Nahua speakers, shoving other peoples off to the margins as Nahuas forged their way south from the Basin of Mexico. We have no indication, however, that they managed or even attempted to overwhelm the Zapotitlán Popolocas or dislodge them from their valley. Perhaps the Zapotitlán Valley was far enough removed from Tehuacán (it is a side valley that enters the Valley of Tehuacán from the west well south of the city of Tehuacán) that the Nahuas did not view it as an impediment or worth conquering. The absence of invasion may also have been due to the willingness of

2. Spencer and Redmond 1997, 25.

FIGURE 6.1 Cerro Cuthá viewed from Zapotitlán Salinas to the west.

the Southern Popolocas, which included those of Cuthá, to align themselves with the Mexica or earlier Nahuas. The Popolocan hilltop fortress of Tepexi el Viejo, north of the Valley of Zapotitlán, was settled around 1300 CE and occupied until the Conquest.[3] Aztecs under Moctezuma I had occupied the fortress by the mid-fifteenth century.[4] These northern Popolocas, who occupied the formerly independent *señoríos* of Tepexi and Tecamachalco, for example, apparently resisted further domination and were subsequently suppressed militarily, conquered by Aztec allies (including Texcoco) in 1503.[5] In this resistance they received no obvious support from their southern Popolocan neighbors, for the northern and southern polities had warred almost continuously for many years, perhaps centuries before the *pax azteca* put an end to such internecine battles. The defensive location of Cerro Cuthá, established long before the rise of Tenochtitlán, was apparently directed toward aggression from the Popolocas of Tepexi rather than enemies from the Mixteca Alta or Oaxaca.

Tepexi el Viejo (not to be confused with Tepexi, the nearby Early Classic site whose residents produced Thin Orange pottery), like Cuthá, was an easily defended

3. Gorenstein 1973, 69. Gorenstein did not identify Tepexi as Popolocan. This was done by Klaus Jäecklein (1979).

4. Castillo Tejero 2002, 177.

5. Gorenstein 1973, 17, 69; Castellón 2006, 53; Jäecklein 1979, 208.

FIGURE 6.2 Zapotitlán Salinas, viewed from Cerro Cuthá.

hilltop site even more elaborately constructed than Cuthá. Artisans there produced highly valued gemstones after the decline of Teotihuacan put an end to demand for Thin Orange pottery, produced a few kilometers away—in a site not defensively oriented. Cuthá, in contrast, produced salt, its inhabitants skilled at production, its influence also widespread. The two precious commodities did not, however, stimulate mutually advantageous trade. Instead, they produced hostility and ongoing strife between the kindred towns. Tepexi resisted Moctezuma's legions; Cuthá folded. The Aztec king established a garrison at Tepexi to assure compliance with his tribute requirements and to "pacify" the inhabitants.

Neither town mounted more than token resistance to Cortés's forces, Cuthá perhaps more than Tepexi. Cortés emptied the Tepexi garrison of Aztec soldiers and required each of the Popolocan towns to relocate, Cuthá to Zapotitlán, Tepexi to the present site of Tepexi de Rodríguez, some 6 kilometers from the ancient site.[6] Spaniards ordered Tepexis to plant mulberry trees and harvest silk cocoons. The project's success was such that from the colonial period until 1876, the town's official name was Tepexi de la Seda (silk).[7] Throughout the same time, Zapotitlán continued

6. Castillo Tejero 2004, 179.

7. Castillo Tejero 2004, 180.

FIGURE 6.3 Retaining wall, Cerro Cuthá. Since the site was excavated in the 1990s, the vegetation has returned with a vengeance.

to produce salt, and to differentiate it from the many other Zapotitláns of central Mexico, it was called Zapotitlán Salinas.

The numerous salt beds of great antiquity in the vicinity of Zapotitlán suggest that the Southern Popolocas supplied salt to Aztec military missions traveling up and down the valleys to the east. The many hundreds of abandoned terraces on surrounding hillsides suggest that they also provided food and other commodities beyond the needs of local consumption and thus escaped the more violent retribution that the Mexica served on less compliant peoples. The valley's name is of Náhuatl origin—the *-tlán* suffix establishes the Nahua imprint—even though the name Zapotitlán may have been derived from the name of King Xhapotl of Cuthá, who fought on the side of the Aztecs against enemies to the west, as Paredes Colín believed.[8] The long-term effects of the early Nahua penetration, including those of the early Aztec conquest, were minimal compared with later, more punitive expeditions from Tenochtitlán to Oaxaca under Ahuízotl and Moctezuma II.

Even before Aztec domination, the Popolocas of Cuthá constituted a *señorío* ruled over by a cacique who appears to have controlled the production and distribution of salt. This cacique would become a puppet of the Mexica or of other peoples

8. Paredes Colín (1910) 1953, 30; *-tlán* is the Náhuatl suffix meaning "place of."

FIGURE 6.4 Salinas (salt pans) near Zapotitlán Salinas. The different colors are produced by different bacteria and algae growing under varying salt concentrations.

before the Aztecs' empire ruled supreme. The Popolocas' sophistication in the complexities of manufacturing salt may have accounted for the Aztecs' willingness to grant Cuthá and environs relative autonomy, but it must have resulted in ongoing hostilities with Tepexi over who would reap the benefits of the salt trade.[9]

The political structure of Cuthá was probably like that of the Mixtec *cacicazgos* of the Late Classic, where towns presided loosely over agrarian communities that would flock to the polis in times of aggression from without. Indeed, Mixtecs may have formed part of the Cuthá population of that period and influenced its political structure.[10] Yet the economic base of the community was different from that of Mixteca Alta and Mixteca Baja. In both, the climate was temperate to chilly, with rainfall sufficient to produce one or two crops of corn each year. The Mixteca Alta was a rich agricultural area, sufficiently populated after the Conquest to support huge teams of workers erecting massive convents and churches for the Dominican priests. In

9. De Benavente (2014, 207–8) notes that Coxcatlán was also known for the production of salt dried in beds. Near San Gabriel Casablanca and San Antonio Nanahuatípam, we saw several operating salt pan facilities. Other sources were in the Mixteca Baja in the Balsas Basin southwest of the Valley of Tehuacán.

10. Castellón 2009, 303–5.

contrast, the Zapotitlán Valley was warm, semiarid to arid, and far less productive of corn, beans, squash, and chiles than the Mixteca. Rainfall at Zapotitlán is just over 400 millimeters annually, insufficient for rain-fed crops, while at Tamazulapan in the Mixteca Alta, rainfall averages 750 millimeters, and the climate is cooler. The Popolocas had to coax production from their stingy terraces to produce a surplus, while it was a much simpler matter in the Mixteca, with its richer soils and more abundant rainfall. What the Zapotitlán Valley had was salt in quantity and quality that made it popular in the Basin of Mexico, lowland Veracruz, and to the south in Oaxaca. Salt production centers are few, and people do value their salt. It is as addicting in the human diet as the most powerful narcotic. It was also an important fixative for dyeing textiles.

After the Spanish pacification of Cuthá, Cortés installed missionaries to solidify the Spanish presence. From there the Spanish forces moved south into the Mixteca Alta and conquered the Chocho towns of Coixtlahuaca and Tamazulapan (now Mixtec), among others. These towns had been allied with Cuthá. (Chochos and Popolocas were apparently close allies politically in addition to their linguistic alignment.[11]) The Crown installed Franciscans in the Tehuacán region and Dominicans in the Mixteca.

The earliest European missionaries to arrive in the Valley of Tehuacán were Franciscan monks. According to their tradition, Popolocas still inhabited the Cerro Cuthá and practiced animal sacrifice to their gods, much to the Brown Robes' dismay. Paredes Colín believed that the friars pressured the residents of Cuthá to relocate from the hilltop to the present site of Zapotitlán, which the Franciscans named San Martín de Zapotitlán. Castellón, however, believes that Cuthá was abandoned and relocated at Zapotitlán well before the arrival of the Franciscans.[12]

Zapotitlán Salinas, the Zapotitlán Valley's principal town, lies only a half hour's drive southwest of Tehuacán, but the contrast between the two locations is notable. The Zapotitlán Valley is more arid in appearance, even though much of it is higher than the Valley of Tehuacán. The soils everywhere appear alkaline (calcareous), and the mosaic of green irrigated fields in Tehuacán Valley to the east is absent. The dominant color of the landscape (except for rainy times in summer) is chalky white to light brown, softened at times by the faint olivine green tint of immense forests of columnar cacti. The channel of the Arroyo Zapotitlán, dry most of the year except below Zapotitlán's sewage outlet, is deeply incised, exceeding 10 meters in depth, reflecting perhaps several hundred years of livestock denuding the soil. Even today the effects of livestock—cattle, burros, and, most numerous, goats—are

11. Castellón 2009, 54.

12. Castellón 2009, 54.

FIGURE 6.5 Erosive cutting in Zapotitlán arroyo.

visible everywhere except where they are fenced out. The whitish lines of terraces, prehistoric and historic, abound, the outlines of ancient milpa sites rising one above the other on the hillsides, reminders of better times for farmers of long ago. Only in the higher reaches of the Sierra Zapotitlán and the foothills of the Sierra Mixteca can one see milpas being farmed. Around the valley bottom, water for irrigation is often unobtainable.

On lofty mesas north of Zapotitlán, pulque producers plant magueys (*Agave salmiana*) and extract aguamiel, the sweet agave juice from which pulque is fermented, much as they have done for over a millennium. They still transport the fermenting beverage to Zapotitlán and other towns, using burros and mules to lug large jugs of pulque in the process of fermentation down the slopes several kilometers to the valley below, where all the available product is consumed by noon each day (see figures 7.27 and 7.28).

The Zapotitlán Valley must have seen more productive agriculture before the arrival of livestock. Traces of the former milpas are everywhere, and ancient *trincheras*—rock walls that edged the terraces and protected milpas from erosion—are evident from afar and up close. Hundreds upon hundreds of faint fan-shaped outlines form gigantic staircases up the many steep hills that line the valley, as well as the Valley of Tehuacán and the Mixteca Alta. Some of the *trincheras* follow contours. Others

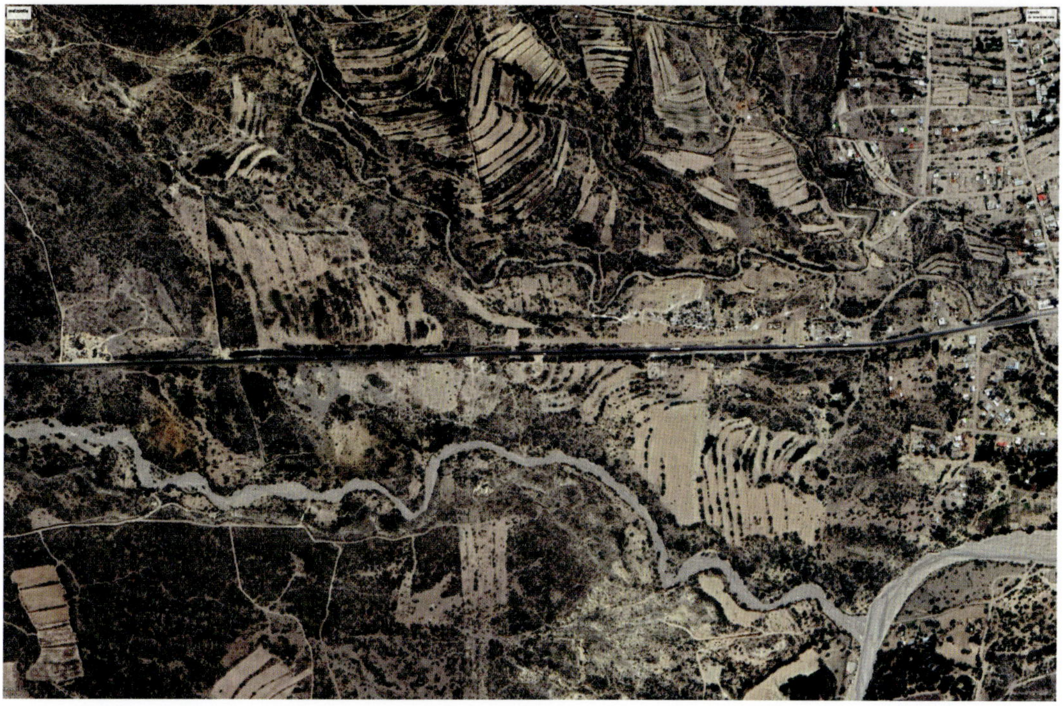

FIGURE 6.6 New, old, and abandoned agricultural terraces in the vicinity of Zapotitlán Salinas. 18.3272 N, -97.4999 W, 1,500 meters. As seen from 4.9 kilometers altitude. Maxxar Technologies and Google Earth.

act as check dams—*lama bordos*—across ditches and drainages. Few of them now support crops, except, perhaps, in those rare years when summer rains are abundant. They hark back to a time when crops supported a population far larger than can now be sustained. Water control systems were widespread and elaborate in the Valley of Tehuacán, extending to most of the Valley of Zapotitlán, for two millennia prior to the Conquest. Paddock argues forcefully that these thousands of formerly productive fields in arid climates bespeak an era when societies could marshal the labor necessary to harvest water and direct it into crop production. The hillsides above the valleys demonstrate their proliferation.[13]

We have no reason to believe that post-Conquest climates were substantially drier than those of pre-Columbian Mesoamerica. What we do know is that diseases of European transmission decimated the Indigenous population of the Americas, killing huge numbers, devastating entire cultures, and rendering the remaining population helpless to organize crop production forces necessary to maintain the vast irrigation and erosion control systems. Populations in Mexico proper and

13. Paddock 1966, 235.

FIGURE 6.7 Ancient roadway among the *tetechos*, Zapotitlán Salinas.

Mesoamerica plummeted to less than 10 percent of their former levels at the same time that Spaniards were forcibly recruiting the few survivors into a new form of labor organization based on Spaniards', not Indigenous, needs.[14] The abandonment of irrigated agrarian systems so evident around Zapotitlán Salinas should come as no surprise. The technology—the complexities involved in crop management—for operating the systems, the thousands of crop management details essential to arid lands agriculture, can easily be lost after one generation of disuse.

The Helia Bravo Hollis Botanical Garden, named after Mexico's most famous cactologist, is located roughly one kilometer northeast of Zapotitlán Salinas. The area has been largely fenced to protect the flora from the predation of goats and is home to one of the densest groves of columnar cacti anywhere, especially the giant *tetecho* (*Neobuxbaumia tetetzo*). The offices and exhibits of the garden sit inconspicuously among a fine mix of succulents, and the grounds extend up an adjacent hillside. They are home to several plants endemic to the valleys. A steady stream of visitors arrives throughout the year, but especially on weekends and in summer, when *chilangos* from Mexico City and *poblanos* from the city of Puebla arrive, much

14. See Reff 1991.

as they probably did ten centuries ago. Guides from the area are readily available to explain the natural history of the gardens. Only a few visitors elect the option of a strenuous three-hour hike up the Cerro Cuthá.

The Cerro Cuthá rises directly east of the botanical gardens. Its summit lies 275 meters above the wash at its base. Viewed from the southwest, its profile resembles the mask of a human face, hence the Popolocan name Cuthá—"The Hill of the Mask." The extensive grouping of monuments atop the cerro is not visible from below, though they were somewhat more visible during archaeological investigations and excavations in the late twentieth century that entailed partial clearing of the summit site. They would have stood out even more imposingly a millennium ago. Now, even after archaeological work and excavation, the thick thornscrub that mantles the hill completely obscures the structures from view except from the hilltop itself or from the air. The growth conceals the many structures that in total occupy roughly ten hectares—the monuments, the palace, the tombs, the stepped platforms, the ball court, the pyramidic buildings of the summit, along with an elaborate network of terraces, retaining walls, stairways, and ramps. In pre-Columbian times, a host of devotees, workers, or slaves would have kept the hilltop cleared and the stairways and roadways to the top free of encroaching vegetation. Without a workforce dedicated to battling the forces of plant growth, within a few years the site will be taken over by the relentless growth of thornscrub, much of it armed with abundant thorns, spines, and clinging vines.

The pathway up the hill, winding and steep, follows an ancient roadway. Midway up, original terraces appear almost out of nowhere. On these artificially constructed flats, villagers built their houses, commoners who maintained the hilltop site and fed and waited on the elite who dwelled above. Fragments of stairways appear sporadically, to guide and add to the wonder the visitor experiences. The original and principal pathway to the summit lies on the southeast side of the cerro. It connects immediately with the ancient road connecting Zapotitlán with the Mixteca Alta and with the less elaborate archaeological site 8 kilometers south, Castillo Rinconada.[15] It too is situated atop a peak, known as Cerro del Castillo Atzingo, which is associated with the Popolocan pueblo of San Juan Atzingo. The ancient route connecting Cuthá and the Mixteca Alta passes the base of this large hill. San Juan Atzingo, which lies a few kilometers east of Castillo Rinconada, happens to be the only town besides Los Reyes Metzontla where the southern dialect of the Popoloca language is still widely spoken. Prehistoric communications between the two sites must have been intimate. Atzingo pueblo and its communal lands jut out into Nahua land.

15. Johnson 1997, 263; Castellón 2009, 30–31.

Walking among the ruins at and near the summit of Cerro Cuthá, one wonders at the lack of maintenance and restoration of such archaeological abundance, the apparent lack of widespread interest in a site exemplifying architectural and engineering sophistication that in other countries would warrant national eminence. How can such a remarkable series of ruins merit only local mention among the pre-Colombian monumental places of Mesoamerica? The answer lies in the many dozens of similar and far more accessible archaeological sites within a radius of 100 kilometers. These include the Mixteca Baja, only 60 kilometers or so south, and the Mixteca Alta, a similar distance southeast, both of which harbor many dozens of spectacular ruins located immediately adjacent to contemporary settlements. Farther southeast lie San José Mogote, Monte Albán, and Mitla amid the archaeologically rich central valleys of Oaxaca. To the north lie multiple vast and rich sites related to the Basin of Mexico. The Cerro Cuthá is but one of hundreds of sites whose cultural and architectural grandeur have been only marginally addressed. Popolocan sites that have seen considerable archaeological activity include Tepexi de Rodríguez, some 40 kilometers north, primarily because of the discovery of its role as a source of Thin Orange pottery, renowned throughout the region in the Early Classic, and Tepexi el Viejo, a nearby well-defended hilltop city with robust trade relations during the Late Classic and Postclassic. East of Tehuacán lies the site of Tehuacán Viejo (La Mesa), a site with large ceremonial and civic structures much more extensive than Cuthá and more readily accessible at the time it was constructed.

Yet the Cuthá site is one of singular importance in the Tehuacán region, an area in which archaeological sites abound. Frederick Peterson, in a reconnaissance for the Tehuacán Project in 1968, identified 370 archaeological sites in the Tehuacán region alone. After what he considered a brief survey and an analysis of ceramics, he concluded that Cuthá was the only fortified site in the Tehuacán region during the early Venta Salada phase, 700–1100 CE, and published a drawn schematic approximation of the site.[16]

The Cerro Cuthá site was described by Nicolás León in an ethnographic study of the Popolocas from 1905. He presented a brief but detailed description of the buildings and related structures at the site. He also remarked that the site was well recognized locally in the nineteenth century, noting that the cacique of Zapotitlán Salinas had ordered protection of the site in 1846.[17]

The most comprehensive study of Cuthá was carried out by Blas Castellón. Castellón was part of the archaeological team undertaking initial excavation and restoration of Tehuacán Viejo (La Mesa) in the mid-1990s. Midway through that project, he

16. The survey was headed by Peterson. MacNeish, Peterson, and Neely 1972, 458–60.

17. León 1905, 22–25.

found the ruins atop the Cerro Cuthá to be of such importance that he dedicated his studies to a greater understanding of the site, surveying the entire site and supervising the excavation of its celebrated Cruciform Tomb. His report was published in 2006.

Castellón's studies reveal a site with its origins in the second century before the Common Era. It gradually developed into a sophisticated community occupied by an organized society. By the Late Classic (650–950 CE), Cuthá had reached its maximum size with a combination of civic and ceremonial functions and a mix of habitations and monuments. Cuthans placed great importance on burial structures, one of which can be seen in the building referred to as the Cruciform Tomb, still accessible though long ago plundered. Remarkable architectural details of the tomb are the lintels and ceiling beams, which appear at first glance to be of resilient lumber. Closer inspection, however, reveals that they are prismatic columnar basalts derived from a mountain some 3 kilometers distant, where they abound in vertical position. They served as lintels without significant alteration from their natural form, the product of rapid and homogeneous cooling of lava flows. The tomb itself was dedicated to a person or personages of elite status, demonstrated by its location among the most prominent structures of the ceremonial/civic sector of the mountaintop. The interior was apparently painted red, the paint probably derived from cochineal, the product of the nopal and the scale insect *Dactylopius coccus*.

The rocks that form the walls delimiting the central plaza are finely carved and carefully emplaced, located with such precision that they remain largely unmoved more than a thousand years later, except where disturbed by looters. They were covered with cream-colored stucco.[18] A peculiar structural component found on Cuthá as well as at Tehuacán Viejo and Coxcatlán is a stone column, often an indicator of the presence of a burial, but doubling in function as a roof or wall support as well.

During the Mesoamerican Late Classic into the Postclassic, that is, between the years 800 CE and 1521 CE, the number of inhabitants living on Cuthá declined, even though the site continued to function until the arrival of Spaniards in 1530. This decrease in population may reflect environmental degradation in the vicinity: Castellón noticed an abundance of abandoned limestone kilns near Cuthá, many of prehistoric origin. Since the production of limestone continued (though on a small scale) until nearly the end of the twentieth century, he consulted with campesinos (peasants) from the Zapotitlán region about their use. According to their re-creation of the technology of lime production, roasting sufficient limestone to produce 50 kilograms (110 lb.) of lime took four days and six workers to stoke the kiln around the clock. In addition, gathering firewood sufficient to produce the 50 kilograms of lime required two to three weeks and the labor of three or four men. He calculates

18. Castellón 2006, 162, 156.

FIGURE 6.8 Burial chamber, Cerro Cuthá. The lintels are single slabs of columnar basalt transported to the top of the cerro from a quarry 3 kilometers distant.

that the firewood necessary to roast the limestone and produce lime for cementing and stuccoing the structures atop Cuthá would have resulted in the general deforestation of the entire region. (Perhaps prior to livestock and deforestation, the rainfall of the Zapotitlán Valley would have been more abundant and plant growth greater.) If we add to the insatiable demands for firewood by the builders of Cuthá the everyday demands for firewood to cook and to heat water, the deforestation must have continued from the beginnings of the occupation of Cuthá in the first century BCE through the end of occupation, perhaps as late as the arrival of Spaniards. Clearly such a sustained campaign of harvesting every burnable twig would have had enormous consequences for the vegetation of the Valley of Zapotitlán and for the local climate. It may also explain the inordinate abundance of columnar cacti, which do not serve well as firewood and whose taxonomic proliferation was already in place long before the advance of organized societies. If we add to the ecological variables the post-Conquest hammering of the vegetation due to unrelenting grazing, especially by goats, the vegetation we see in the Valley of Zapotitlán must be utterly distinct—shorter, sparser, and poorer in species—from its appearance prior to the building of the great complex atop the Cerro Cuthá.

Standing on the summit, we spotted several salinas glimmering in the distance, each supporting dozens of evaporation pans or ponds, each pan with a slightly different hue, a phenomenon caused by algae growing at varying salt concentrations. The presence of so many salt pans over such an extended period suggests that the Cuthá site had economic and strategic reasons for its existence in addition to any religious or ceremonial importance. A satellite view reveals the importance of the salt beds to the Zapotitlán region. Over the centuries, beginning well before the time of Conquest, residents have exploited multiple localities with hundreds of salt pans or beds, many of which are still in production. According to Castellón, pre-Columbian technology at Cuthá involved heating the concentrated brines produced through solar evaporation in specialized vessels at a relatively low temperature until moisture had evaporated. Thus a salt highly valued for human consumption was produced.[19] The product was molded into blocks and made available for export and, in all probability, tribute.[20] Cuthans undoubtedly utilized solar drying pans as well. These were small rectangular ponds, meticulously maintained. They required considerably more time to produce salt because rain diluted the concentrate, high humidity delayed the drying, and the pans required ongoing maintenance to level the terrain, create an impermeable layer, and protect them from erosion damage from runoff above the pans.

19. Castellón 2009, 75–80.

20. For a description of pre-Columbian salt technology, see Williams 2016, 45ff.

FIGURE 6.9 Salt pans based on pre-Columbian structures near Zapotitlán Salinas.

In the pre-Columbian Mexico City basin, cottage salt production was a part-time activity in some areas, while entire towns were devoted to full-time production in others.[21] Salt producers on a domestic scale labored during the dry season, when the lake levels were low and low humidity permitted rapid drying, to produce salt from the muck left behind by the receding lakes. During the wet season, they tended their milpas. In Zapotitlán it is quite possible that salt workers devoted full time to salt extraction, using solar production during the dry season and the more labor-intensive applied heat during the rainy season. This implies, of course, that the region could produce an agricultural surplus that would support full-time salt producers or that salt could be traded for commodities essential to the community's survival.

Spaniards introduced a rationalized salt pan technology. It required more precise leveling and sealing of the salt beds and relied entirely on solar energy to evaporate the water from brine. This technique permits mass production of salt, an important reagent and reducing agent in the *beneficio de patio* process for producing silver from lower-grade ores, in which brine was mixed with mercury. The new technology

21. See De León 2010. Iztacalco, Xocotitlán, and Nexquipayac, among other towns situated around Lake Texcoco, were dedicated to part- or full-time salt production. As of 2014 the seasonal production of salt continued on Lake Texcoco. The proposed construction of the Mexico City airport, now cancelled, would have eliminated the last source of the specialized salt called *tequesquite* for the few remaining producers in the Basin of Mexico. Parsons 1989.

FIGURE 6.10 Salt pans at Las Ventas Indigenous Community near Texcala, Puebla. The installation is communally owned and operated.

was introduced early in the colonial period and satisfied an increased demand for salt, which colonists also required in large quantities for livestock consumption and to produce leather. The beds, carefully excavated and leveled, are vulnerable to destruction by rains and flooding, even relatively light amounts, so their maintenance required concentrated labor and vigilance as they do today, especially during the summer rainy season. Some of the beds are located perilously close to unstable banks of adjacent arroyos. Still, the industry flourished, and what the Zapotitlán Valley lacked (or lost) in agricultural production, it made up for in salt. It may have been the closest source of large-scale production of high-value salt production over a wide geographic area, perhaps including Oaxaca. Exports from the salinas reached as far north as Pachuca, Hidalgo, north of Mexico City, and as far south as Guatemala.[22] Other centers of production existed, but the perceived quality of Zapotitlán salt even then produced high demand for the product. Today, the *panes* (salt beds) of Zapotitlán de la Salinas continue to yield various salts. Control over the beds is a matter of controversy and has been so for many centuries. The Crown ruled that only Indigenous people from Zapotitlán could work the salinas.[23]

22. Williams 2016.

23. Paredes Colín (1910) 1953. The Indigenous community of Las Ventas, southeast of Texcala, controls its own salinas.

Salt was also produced in pre-Columbian times in the region of Coxcatlán in the Valley of Tehuacán, including a source on Cerro Petlanco, but these have fallen out of use. Salt seeps can still be found throughout the region, but those near Zapotitlán seem the most reliable, and those trickling from the Cerro Cuthá and nearby are the most productive of all, hence the name Zapotitlán Salinas. Fifteen different seeps are in production when climatic conditions permit, several of them visible from the summit—when it has been cleared of vegetation, that is.

Presa Purrón—The Purrón Dam

About 40 kilometers southeast of Cerro Cuthá, the most important hydraulic structure known from Mesoamerica rises inconspicuously from a desert dry wash. Compared with the monumental temples and pyramids of Monte Albán or the massive and complex walls and ball court of Cerro Quiotepec, the site does not prepare the visitor for anything unusual. Aside from the imposing landscape of whitish cliffs of the Oaxaca Fault in the distance, the ruins of Presa Purrón seem downright nondescript. No road signs inform the traveler of its existence. The dirt roadway leading to it is best negotiated in a high-clearance vehicle. No one lives nearby. The remains form a bluff at the edge of an arroyo that is usually dry. They appear as a gash in the hillside, surrounded by thornscrub trampled by cattle and savaged by goats, thick with thorny and spiny desert plants—cacti, acacias, catclaws, and mesquites. The dam works blend so subtly with the arid tan and white soils and substrate, as well as with the water-polished stones of the arroyo conglomerates, that one can easily pass by the site oblivious to the great structure that once housed the largest dam in pre-Columbian Mesoamerica, perhaps in all the Americas.[24] On closer inspection, the gap reveals a superstructure consisting of untold thousands of cut stones and bricks fashioned into buttresses and ramparts.

Once we noticed the rockwork towering above the arroyo bottom, it became clear that all these stones and bricks contributed to a massive structure that had once crossed the arroyo and extended far to the north, where it joined the base of a steep hill nearly a half kilometer away. Recent studies demonstrate that at its fullest development, the structure was 24 meters high, 106 meters wide, and 400 meters long.[25] Only by inspecting the site on the ground and viewing the massive buttresses incorporated into the structure can one imagine the sophisticated engineering involved; the colossal amounts of fill, stone, and adobes required to construct the dam;

24. Dams appeared throughout pre-Columbian Mesoamerica, most of them much later, but none on the scale of Presa Purrón.

25. Woodbury and Neely 1972; Aiuvalasit, Neely, and Bateman 2010, 1208.

FIGURE 6.11 Presa Purrón (Purrón Dam), southern terminus, near Tilapa, Puebla.

and the extraordinary maintenance required to patch leaks, maintain canals and spillways, remove the accumulating sediments, and preserve the overall physical integrity of the dam.

Even so, in the long run, the dam failed, probably not because of faulty engineering, but because inevitable buildups of sediments kept requiring a larger and larger structure, and because of economic change. Now, as a result of more than two millennia of silt buildup, pervasive thornscrub, the steepness of the cliffs along the southern side, and the trampling of millions of livestock hooves, a comprehensive view of the entire dam is elusive, which was undoubtedly a factor in the delay in systematic studies of the structure to fill out the archaeological record. It was not until the early 1960s that archaeologists became aware of the structure.[26] As late as 1990 William Doolittle suspected that the dam was strictly a flood control structure, since diversion canals had not yet been identified.[27] Only in the late 1990s did researchers locate and study a series of canals that conveyed impounded water to the parched fields of the lower valley. They also identified a major canal, christened the Santa María, that extends for 1.4 kilometers, well above the highest watermark of the dam on the northern margin, most likely to carry water to fields while the arroyo was

26. Woodbury and Neely 1972, 82. The authors credit Richard MacNeish with the realization that the structure was in fact a dam.

27. Doolittle 1990, 28.

FIGURE 6.12 Santa María Canal, Presa Purrón, located above the dam's high-water point.

running.[28] Ejidatarios of San José Tilapa were quite willing to point out the canal's outline to us and guide us through much of its course.

Presa Purrón might not have existed if the technicians of the advanced civilization in Teopantecuanitlán had not experimented with water control measures during the Formative period, well before the construction of Presa Purrón. There, about three thousand years ago, in the mountains of the state of Guerrero, roughly 200 kilometers west of Lencho Diego arroyo, Olmecs or related peoples designed and operated the first known dam in the Americas, which started a long-term series of dam constructions in pre-Hispanic Mesoamerica.[29] The discovery of a technologically sophisticated culture so early in the development of Mesoamerican cultures called into question many established theories and extended well back into antiquity our understanding of the accomplishments of civilizations in the Americas well before the beginning of the Common Era.

Presa Purrón (also known as Presa Mequitongo) is located just northwest of the small town of Tilapa in extreme southern Puebla, some 20 kilometers from the

28. Aiuvalasit, Neely, and Bateman 2010.

29. Scarborough 2000.

Coxcatlán Cave made famous by MacNeish and his associates. The dam structure lies within the boundaries of lands belonging to an Indigenous community in which many residents still speak Náhuatl. Most men work in sugarcane fields and run livestock on communally owned lands. The community elects a commissar who oversees access to the dam. Most residents seem to be familiar with the site and are aware that a great dam once lay across the Lencho Diego arroyo. Villagers have run their cows, burros, and goats over the site for centuries and have long known of the untold millions of stones and adobes emplaced at the site. Whether their take on the structure was part of the archaeological discovery does not appear to have been recorded.

The steep hill that anchors the northern end of the dam is known locally as the Cerro Lencho Diego, named for a famous bandit (Lencho is short for Lorenzo). Residents hold the hill in high regard, endowing it with mystical properties. Indeed, a cave carved into the hillside, said to have served as Lencho Diego's stronghold, has been ransacked by treasure seekers and adopted as a refuge by bats. The hill, composed largely of gypsum, is riddled with holes and tiny passages. From the summit one can gain a passable view of the dam and detect the acrid odor of bat guano seeping from cracks and vents on the hilltop, adding to the hill's mystique.

Lencho Diego arroyo originates in the Sierra Chichiltepec to the northeast, part of the massive range to the east that dominates the Valley of Tehuacán. Its catchment area is about 3,000 hectares (7,500 acres). If we can imagine covering the whole basin upstream with a large plastic sheet and ignore evaporation, rainfall would run downslope from small crevices into rivulets, brooks, gullies, small creeks, and drain into the mainstream. Being conservative and taking an average of 650 millimeters of annual precipitation across the catchment area (meaning 0.65 cubic meters on each surface meter), in one year this small basin would have collected almost 20 million cubic meters. Of course, there are huge losses to evaporation, accumulation and usage by plants, and soil infiltration. These, according to hydrologists, account for about 70 percent of total precipitation, so roughly 2.5–6 million cubic meters would have been available for storage and irrigation each year.[30]

The arroyo has long since cut through the northern edge of the dam. A smaller arroyo with a lower gradient enters the dam site to the south and has made the most visible breach of the structure, easily seen from Cerro Las Escaleras, which forms the southern rampart of the dam. This smaller drainage joins the larger Lencho Diego arroyo just downstream from the dam. Prior to the dam's last period of construction, the engineers realized that constructing a 400-meter-long structure would capture the waters of both drainages. They clearly selected one of the best sites in the entire region.

30. Woodbury and Neely 1972; Aiuvalasit, Neely, and Bateman 2010; Terzaghi 1986, 219–20.

FIGURE 6.13 Close-up of dam structures, Presa Purrón.

Determining the dam's age through its several phases has proved a difficult challenge. The earliest structure was modest, according to the initial archaeological excavations, measuring roughly 2.8 meters high, 6 meters wide, and no more than 170 meters long. Since the southern arroyo is nearly 10 meters lower than the Lencho Diego arroyo to the north, there was no need for this early dam to extend the full width of the arroyo complex. Using the remains of ceramics and stratigraphy, the first archaeological excavators estimated that this earliest dam was constructed between 750 and 600 BCE, a time frame upheld by recent radiometric studies. In other words, dam building began on the arroyo more than 2,500 years ago.

As population in the region expanded and sedimentation rapidly filled the dam's storage area, the need for greater water storage became apparent.[31] Rainfall in the lowest elevation reaches of the Valley of Tehuacán is generally little more than 400 millimeters annually, hardly enough to raise a single crop of corn, and then only if the rains fall in a timely fashion. Higher elevations (including the dam site) receive somewhat higher rainfall, but nowhere in the valley bottoms is rainfall sufficient for dependable corn production. Irrigation vastly augments corn harvests. Studies at the Purrón Cave, not far from the dam site, also show a strong correlation between

31. Woodbury and Neely (1972) noted that a coffer dam to keep silting at bay is still visible upstream.

the appearance of orchard tree crops from the region and the operation of the dam.[32] Plant remains in the cave's stratigraphy show that the area must have contained irrigated orchards during the Santa María and early Palo Blanco phases—roughly 900 BCE–300 CE. Plant parts of (among others) coyol palm fruits (*Acrocomia aculeata*), *ciruela* (*Spondias mombin*), *chupandilla* (*Cyrtocarpa procera*), *cosahuico* (*Sideroxylon tepicense*), white zapote (*Casimiroa edulis*), and, most important, avocado (*Persea americana*) appear suddenly in the archaeological sequence. Fruits of these trees are absent prior to the dates of the dam's construction. The same fruit residues in roughly the same abundance were discovered in the nearby Coxcatlán Cave. In addition to the abundance of the plant parts, the size of the seeds increased notably during those sequence years, which coincides remarkably with years in which the dam was in operation. When the dam ceased to function early in the Common Era, abundance of the fruit remnants decreased. C. Earle Smith concluded from this data that irrigation from the dam was responsible for dramatic increases in plant availability, size, and vigor.[33]

The avocado was probably domesticated in the vicinity of Coxcatlán Cave, which means that its early cultivars were most likely developed from the Purrón irrigation district.[34] Wild or semidomesticated avocados are still available in the public markets of Tehuacán and smaller cities in the region. These fruits are the size of a small lemon, far smaller than the commercial avocados now available. The pulp of the wild fruits is protected by a thin, leathery skin that is usually eaten or spat out. Smith suggests that the process of domestication of the tree was necessarily a lengthy one and attributes a long history of cultivation to the farmers of the Valley of Tehuacán, and we must also submit attribution to the engineers who designed and oversaw construction of Purrón Dam and other smaller hydraulic works.

The dam was expanded on at least three different occasions. The final phase was completed sometime prior to 300 CE, after which time it appears the dam and irrigation works were slowly abandoned. (The gradual abandonment appears to coincide with the Zapotec withdrawal from Cerro de Quiotepec.) Without an organized workforce to maintain the dam works, the arroyos breached it and slowly ate away at both ends, rendering it useless as a dam. In the Late Classic period, perhaps as late as 1100 CE, residents constructed a pyramid atop the remains of the dam, a use unrelated to water management.[35] In other words, the dam structure stood breached and unused for impounding water for at least 1,000 years before the arrival of Spaniards.

32. C. E. Smith 1967, 230.

33. C. E. Smith 1967, 230.

34. Galindo-Tovar et al. 2007.

35. Woodbury and Neely 1972, 92–94.

And yet a large population would have been necessary to construct the dam, and a large agricultural workforce would have been necessary to support the caloric needs of the workers. The population must have considered the enormous work input worth the cost, since the final structure was carefully engineered, and huge numbers of carved stones carefully emplaced, while skillfully designed buttresses strengthened and reinforced the structure. Both were essential to the structural integrity and operation of the dam. (A dam is an energy and resources machine, albeit a large, cumbersome one.)

Richard Woodbury and James Neely estimated that the final phase of construction required nearly 200,000 cubic meters of fill and would have required the labor of several thousand workers over a year's time to complete the project. Clearly a well-organized society with sophisticated managerial capability and well-developed understanding of principles of civil engineering and hydraulics was a requisite to marshal and organize the workforce and construct a lasting dam. And the water impounded by the facility must have provided sufficient irrigation potential (perhaps sufficient for roughly 675 hectares) to provide for a sizable population and justify the enormous effort required to build the dam.

Who were the people responsible for constructing and operating the sequence of dams over a period of nearly a thousand years? No one has ventured a guess, and any surmise is speculative. The most likely candidates, however, would be Popolocas from the northwest, whose territory extended south of Coxcatlán in Classic times (some 1,000–1,500 years after initial construction of the dam).[36] Nahuas, present Indigenous residents of rural portions of the Valley of Tehuacán, did not arrive until the eight century CE at the earliest (perhaps not until the twelfth century). They may have been responsible for the temple constructed on the dam in the twelfth century, but there is no evidence that they were present in the region at the time of the dam's construction or that they exploited its water storage for irrigation. Zapotecs had conquered Quiotepec and the Cuicatlán Valley by the Late Formative (Monte Albán I), around 300 BCE, which coincides with the final two phases of dam construction, but evidence suggests they limited their northward expansion to the confluence of the Río Salado and Río Grande, some 30 kilometers south. It seems oddly coincidental that the demise of the dam or the society that built and maintained it corresponded roughly with the withdrawal of Zapotecs from the La Cañada between 200 and 300 CE. It is just possible that Zapotecs took advantage of food commodities, necessary to provide for their garrisons, that became available thanks to the expanded acreage irrigated by waters of the dam. When the Zapotec demand diminished at the time of their withdrawal back to Monte Albán, it seems likely that

36. Gámez 2003, fig. 2.

the demand for crops and water languished and with it the support of the workforce required to maintain the hydraulic culture of the lower valley of Tehuacán. When the Zapotecs departed, the orchards of Lencho Diego arroyo dried up, or, perhaps, they departed *when* the orchards dried up. Perhaps the soldiers staffing the garrison of Quiotepec had also benefited from the intensive agriculture of Purrón, and their departure sealed its doom, or vice versa. When a major consumer departs an area, the supportive industries suffer, as is graphically evident along stretches of U.S. Route 66 bypassed by interstate highways in the late 1960s. Or perhaps drought or silting dried up the dam's ability to irrigate, and the Zapotec fortress lacked sufficient food to support its function. Diseases may have decimated the workforce and affected the garrison as well. Whatever the causal sequence, the dam lost its importance, and the Zapotec troops departed at roughly the same time.

Other peoples in the region could well have been involved in the construction and operation of the dam. Little is known of the ethnicity of the builders and residents at and around the sites of Tecomavaca or Los Cues, both of which lie within 25 kilometers of the dam, but the labor required to construct the civic and ceremonial complexes at those settlements would have been considerable.[37] Bandelier noted in the late nineteenth century that the inhabitants of both towns were of Mazatec origin, but we have not uncovered further information about the people who built the sites or about the age of the sites.[38] Mazatecs have since been assimilated or departed to the east.

The very existence of the Purrón Dam implies the vigorous activity and complex social relations of an organized society trading with others. Travelers had to carry all supplies on their backs, so they needed local food, and farmers were only too pleased to provide—for the right price, of course. The dam and the more than six hundred irrigated hectares (any of which could have been double cropped with corn) would provide for a large population but also for the hungry merchants and their entourages that came and went in all directions. But only an organized society could marshal the knowledge and energy to construct the enormous structure. Only cadres of sophisticated engineers, water experts, builders, and overseers could have conceived and carried out the dam's design. The corridor connecting Mexico with Mesoamerica must have been a busy one, even 2,500 years ago.

Even as the dam ceased to function, however (gradually, we must assume, but perhaps as the result of a single catastrophic breach due to a massive storm in the watershed), brisk trade continued through the valleys. At least some of it appears

37. *Cues* in Náhuatl mean "Hills in the form or figure of temples" or "Round hills where temples have been built." Cue is a well-known surname in Oaxaca, including that of a recent state governor. Bandelier (1884, 266) describes the nearby archaeological sites of these two villages.

38. Bandelier 1884, 266–68.

to have been carried on independent of the productive relations within the valleys themselves. A study of obsidian tool production sites within the Valley of Tehuacán revealed that during the Late Formative into the Early and Middle Classic periods, the sites were curiously independent of local social connections. Shops that had no local outlet produced obsidian artifacts and perhaps pottery. The obsidian arrived from sources to the north, was fashioned into rough or finished tools, and was then shipped on to points south. The products seem to have served exclusively clients of Teotihuacan commerce, extending from Olmec country in the lowlands of Veracruz (via Quiotepec) or the Maya city of Kaminaljuyú, Guatemala.[39] The production shops were not associated with any polities within the valleys but functioned independently. It was as though they were pre-Columbian *maquiladoras* (assembly plants), industries that may have relied on local labor but whose products did not circulate locally. The brisk trade implied by these factories also implies the need for a reliable source of food for traders. It seems that agriculture in the Tehuacán Valley would have been essential to the maintenance of this traffic. The producers must have missed the bounty of fields irrigated by the Presa Purrón as they slowly dried up in the Early to Middle Classic.

Cerro de Quiotepec

If Presa Purrón lay abandoned and unrecognized for centuries because of its camouflaged existence in an ostensibly nondescript dry wash setting, the fortress called the Cerro Quiotepec, some 25 kilometers south, stands out as a conspicuous structure in a charismatic location. Zapotecs chose carefully and well when selecting strategic sites to solidify their imperial expansion and preserve their territorial gains. The hilltop site of Monte Albán, a few kilometers southwest of the city of Oaxaca, combines a readily defensible site with the lofty heights appropriate to an imperial power presiding over vast and fertile agricultural valleys. It is also within eyesight of the Cerro Atzompa, some 4 kilometers north, a well-fortified auxiliary site. From Atzompa one can see another hilltop site 20 kilometers north, at Suchilquitongo, and so on, line-of-sight structures quite possibly existed all the way down Cañada Chica to La Cañada and Cerro Quiotepec (see figure 3.1).

Monte Albán constituted a combined military, spiritual, and administrative capital intended to control and to instill awe in commoners and enemies alike. Over the centuries the elite of Monte Albán buffered their stronghold with homes and businesses on terraces below, so that attackers would have to fight their way up, over

39. Drennan, Fitzgibbons, and Dehn 2000, 178, fig. 8.2, 184–88.

FIGURE 6.14 Cerro Quiotepec summit.

walls, and through streets and alleys before reaching the mountaintop site, where the arrogant rulers and their retinues held court. As long as the leaders looked after the public interest, homeowners and merchants had a vested interest in fending off raiders and thieves, as the lords knew.

The hilltop site of the Quiotepec fortress is as obvious a choice as Monte Albán for military purposes, though the latter was not constructed exclusively for defense. The Quiotepec site provided Monte Albán an early bulwark against potential attackers from the north, south, and east (no militarily significant competitors from the west have yet been established) but also enabled the Zapotecs to monitor and control trade between the Río Grande and the Río Salado, that is, between the valleys of Cuicatlán and Tehuacán and from the Río Papaloapan basin. The site may also have proved intimidating for Mixtecs from the west, but as we have noted, their presence in the archaeological record is minimal.

Administration of the site, as remote from the nearest city as any of Mexico's pre-Columbian monuments, now lies with Mexico's Instituto Nacional de Antropología e Historia (INAH). Day-to-day visitation permits, however, are administered by officials of the community of Quiotepec, which lies perhaps 2 kilometers from the site, across the Río Grande, on the east bank. The site is not accessible by motor vehicle. Special arrangements with local ranchers and a local guide are required. Villagers serve as guides to the site, and INAH employs them to clear the site and maintain it free of brush and easily removable artifacts, perhaps to the detriment of the preservation of the ruins' integrity. The river crossing that villagers use to gain access to the site is probably the same ford used for millennia by travelers and soldiers. During times of high water in the twentieth century, the only access to and from the town was across the railroad bridge, now abandoned, which for several years was the only practical vehicular access to Cerro Quiotepec as well. A large highway bridge connecting Quiotepec to Highway 135 was washed out in the massive floods that scoured the region and the water courses in 2011 and was not replaced until 2014.

The fortress and related structures occupy 45 hectares (110 acres).[40] The complex has been only partially excavated, and restoration has been minimal but sufficient to demonstrate the huge resources Zapotecs invested in construction. Massive walls 10 meters high on the summit rise from the edge of sheer cliffs on the northwest side, while a substantial wall constructed from cut stone and mortar and often faced with stucco covers much of the remainder of the periphery. Two large plazas, one containing a platform mound 60 meters by 55 meters and an extensive residential complex, form the very apex of the hill. The site is laced with tombs lined with murals and was once packed with a trove of treasured items that accompanied the deceased. Looters have long since pillaged most of the artifacts. A ball court 65 meters long occupies one of the few flat areas (artificially flattened) on the cerro. The ball court suggests that the site also functioned as a ceremonial center. Roughly thirty residences accommodated the high military officials who directed the site's activities. Spencer and Redmond estimate that 110–115 residences were part of the Quiotepec complex.[41] Overall, the site functioned as both Monte Albán's northernmost ersatz custom house and a foreign military base and garrison.

During the Zapotecan occupation, roughly 300 BCE to 300 CE, in addition to the staff of the garrison, as many as two thousand people resided in the nearby area at a total of seven sites.[42] A large proportion of this population would have consisted of the agricultural workforce required to provide for the largely military population and their families. Provisioning the site with water, carried in jugs up to the summit more than 200 meters above the river, would have required a sizable workforce.

Viewing the surroundings from the summit of Cerro Quiotepec, one can see that a relatively small military force could stop or slow the advance of a much larger army arriving from any direction, or intercept merchant convoys.[43] The hilltop location affords clear views many kilometers upstream in La Cañada. When observation

40. Spencer and Redmond 1997, 12.

41. Spencer and Redmond 1997, 533.

42. Spencer and Redmond 1997, 601.

43. Hassig (1992, 35) notes that as a rule, successful assaults of fortifications require three attackers for each defender. In the case of Cerro Quiotepec, the defenders would have enjoyed an even greater advantage.

FIGURE 6.15 Retaining walls surrounding the summit, Cerro Quiotepec. Note the fine blockwork.

FIGURE 6.16 Corner wall of civic or religious structure, Cerro Quiotepec. Note remnants of the decorative frieze.

FIGURE 6.17 Valley of Tehuacán, looking north up the Río Salado from Cerro Quiotepec, with flowering *Agave quiotepecensis*.

FIGURE 6.18 La Cañada, looking south from Cerro Quiotepec. The structure sits atop the wall pictured in figure 6.16. The pass, Paso Quiotepec, which connects the Cuicatlán and Tehuacán Valleys, is located on the upper right side of the photograph.

points on the summit and adjoining hilltops are added to the cerro itself, approaches of military units of any significance from the north, south, or east would have become easily observable. The steep mountains to the west would prevent any mass attack on the fortress from that direction. Part of the complex seen clearly from the summit features a broad (200 m by 210 m) plaza, flanked by long platform mounds 6 meters high. Any southbound traveler hoping to ford the river would have to climb several steps and pass between the platforms, then descend steps on the far side to the crossing. The same gauntlet would face any northbound traveler, who would have arrived sopping wet, having just forded the river, which runs nearly waist deep even at times of low flow. These structures demonstrate how easily a small force could have controlled north–south and east–west transit, as Zapotecs discovered when they assumed tight control over all commerce from the north and heading north.[44]

From the very summit of the cerro, a view downstream into the profound canyon plunging eastward would reveal all but tiny bands of adventurers, merchants, or troops. The peculiar acoustics of the site also make sounds from the canyon downstream audible at the very summit, roughly 250 meters higher in elevation from the

44. Redmond 1983, 99; Spencer and Redmond 1997, 601.

FIGURE 6.19 The ford of the Río Grande at Quiotepec. Alberto Búrquez is the rider.

point where westward travelers would have first seen the fortification come into view. The hilltop site is atop steep slopes, and the massive walls constructed at its apex could enable defenders to resist attack for extended periods. The Río Santo Domingo was the gateway to the Gulf of Mexico from La Cañada. Whoever controlled the Quiotepec site also controlled trade to Teotihuacan, Cholula, and Tehuacán to the north, to Oaxaca and the Mixteca to the south, and to Tuxtepec, the Gulf coast, and the Olmecs to the east. This last route is especially important because the Río Papaloapan allowed easy traveling by water across otherwise impenetrable tropical lowland forests and coastal mangroves. The Papaloapan also allowed access to the sea, where a string of coastal trading places were located. Olmecs would have known the Cerro Quiotepec well, if only as a landmark.

Between 200 BCE and 100 CE Zapotecs achieved an unprecedented level of political organization—the formation of a full-blown state—at Monte Albán, the firstborn in the Oaxaca region.[45] In keeping with this expansion and their rapidly evolving administrative and class structure, they developed the imperial policy of extending and protecting their borders and their trade. Early in this period, around

45. Marcus and Flannery 2000, 381. Mixtecs had already achieved a high degree of technical advancement but did not construct city-states.

FIGURE 6.20 Looking downstream into the Santo Domingo canyon from the wall, Cerro Quiotepec. Note the myriad milpas on the Sierra Mazateca.

200 BCE or somewhat earlier, their troops invaded La Cañada, executed many of the previous inhabitants, shoved the others aside, and shored up the defenses at Quiotepec, establishing colonial rule that included control over intervalley traffic and the exaction of tribute. That is what colonies are for.

The ethnic identity of the peoples who were overwhelmed by the Zapotec invasion is unclear. The logical candidate is Cuicatecs, but the victims of the Zapotec onslaught could also have been Mazatecs or Mixtecs, since both peoples are known to have shared power with Cuicatecs in La Cañada. When the Zapotecs did retreat from or abandon Quiotepec half a millennium later, Cuicatecs or Mazatecs quite probably filtered back into the site, planting their corn on the river terraces below the mountainous terrain. From the higher points of the cerro, one can still see large nearly pure expanses of boat-thorned acacia (*Vachellia campechiana*), small trees of uniform size, a pioneer plant in natural succession following soil disturbance. This sort of vegetation cover is indicative of previous sustained agricultural use. From the train station of Quiotepec, the now-unused railroad tracks follow the riverbanks in a gentle downward grade on what were probably the pre-Hispanic foundations of a diversion canal to irrigate this roughly forty acre "milpa." Other ancient diversion canals (*apantles*) are common throughout the region. Some are still in use (see figure 4.12).[46]

46. C. E. Smith 1965a, 68ff.

The summit of the Cerro de Quiotepec sits only about 700 meters (2,297 ft.) above sea level and around 250 meters above the junction of the two rivers at its eastern base. Quiotepec's climate is one of the hottest in southern Mexico—temperatures in May reach 46°C (115°F), unlike the springlike conditions at Monte Albán. Any duty for Zapotecs stationed at the fortress must have constituted a considerable imposition. At the same time, the lords of Monte Albán must have considered the hostile climate tolerable and the huge expenditure of resources for constructing and maintaining the fortress and environs a worthwhile investment. Whether they were extracting import or export duties, restricting immigration, or protecting borders, the Zapotecs went to great lengths to construct the customs installation and garrison and maintain it for several centuries.

The summit fortifications not only overlooked valleys to the north, south, and east, but allowed a small force of soldiery to control access from all directions. Tehuacán Valley and the Río Grande narrow considerably near their confluence, and trails are relegated to hillsides, but the pass connecting the two, Paso Quiotepec, requires an elevation gain of little more than 40 meters, so it is the logical point of passage between the two valleys and directly connects to the Basin of Mexico to the north and west and to southern Mesoamerica and its cultures to the south and east. The Río Grande, flowing generally to the north, is forced abruptly eastward by the Cerro Quiotepec, while the Río Salado, flowing generally southeastward, is directed to the east by the cerro as well, but also by the uplift of the Sierra Mixteca.[47] For travelers to bypass the confluence and the Cerro de Quiotepec with its watchtowers would require an arduous detour, no matter the direction of travel. Solitary travelers could climb and descend the slopes clandestinely and thus evade the watchful eyes of the Zapotecs. Larger groups could only do so with a great expenditure of energy. Detours into the Mixteca Alta to the west involved more mountainous terrain and crossing the lands of several different cultural groups, each probably suspicious of outsiders and each with their own set of questions to ask of travelers.

What we know about the fortress is intricately enmeshed in what we know about the prehistory of La Cañada, Cañada Chica, and the lower Río Grande canyon country. The archaeological record has been mostly the result of work by Charles Spencer and Elsa Redmond, which they undertook in 1977. To them and their archaeological team, we are indebted not merely for their painstaking work and reconstruction of the region's ancient history, but for their perseverance through the stifling heat and humidity of much of the year in La Cañada. The sides and terraces of canyon bottom, where it is not rendered lush by the waters of the river or by irrigation, are vegetated

47. The range north of the Santo Domingo breach is often referred to simply as the Sierra Zongolica. Ten kilometers or so north, near the town of Calipan, the Río Comulco joins the Río Salado from the east. Some geographers refer to that section of the sierras as the Sierra Mazateca. Its highest peak is about 3,270 meters elevation.

mostly with dense growth of acacias, catclaws, mesquites, nettles, palo verdes, numerous species of cacti, and, when the summer rains arrive, a jumble of vines. All of these conspire to make passage through the thorny *monte*, as this vegetation is called locally, a challenge and a hazard to one's skin. Farther up the sides of the canyon, the vegetation merges into tropical deciduous forest, even more dense though not quite as thorny, with the addition of several species of *Bursera* and the prominent *pochote* (kapok) tree, *Ceiba aesculifolia*.

While attempting to avoid the thorny, spiny, and sticky underbrush that flourish in the lower portions of the canyon, one usually experiences an infestation of chiggers (perhaps introduced by cattle) and no-see-ums during the day and mosquitoes at night. The thorny desert scrub the archaeological teams were forced to remove were also a formidable barrier to their archaeological reconstruction and research at the Llano Perdido site at La Coyotera, not far from the town of Santiago Dominguillo. The growth of desert vegetation had become dismayingly dense in the two millennia since the settlement was abandoned and excavation begun under the direction of Spencer and Redmond. At the conclusion of their work, they ordered the site backfilled. Llano Perdido has become a pea gravel–covered soccer court for local youth, one of the few nonagricultural flat spots in the region.

Spencer and Redmond were attracted to La Cañada by its strategic location, its abundant waters, and its rich history in producing tropical foods. Even more influential in their interest, however, was the pronouncement in 1976 by archaeologist Joyce Marcus that an inscribed stone panel on Building J of Monte Albán (also called the Observatory) appeared to commemorate or celebrate the conquest of Cuicatlán by the Zapotecs. Researchers had already established the date of construction of Building J as prior to 200 CE, so the conquest celebration had to have taken place prior to that time, most likely during a period in the Monte Albán prehistory known as the Monte Albán I.

As a result of their investigations in La Cañada, Spencer and Redmond divided the prehistory into four periods: Perdido phase (600–300 BCE), Lomas phase (300 BCE–200 CE), Trujano phase (200 CE–1000 CE), and Iglesia Vieja phase (1000 CE to Conquest in 1525). They ascertained that people of unknown identity had developed a rich culture in La Cañada and Cañada Chica extending as far back as 600 BCE. These people practiced ancestor worship, lived in communal residences, and experienced a class society in which the elite collected prestigious artifacts, including shells from the Pacific Ocean, obsidian, magnetite stone for inlaying in shell ornaments, and imported pottery.[48] The elite of this pre-Zapotec society were also afforded burials accompanied by high-quality items, primarily fine

48. Spencer and Redmond 1997, 461–65.

pottery. These accompaniments marked their high status during their lifetimes, and custom permitted them to take their treasures with them. Commoners were buried with few or no ornamental objects.

The largest Perdido site excavated was at La Coyotera, only a few meters above the floodway of the Río de las Vueltas, a site Spencer and Redmond labeled Llano Perdido. The results of the investigation suggest that the farmers of Llano Perdido used simple field irrigation, with water diverted directly from watercourses (in this case, the Río de las Vueltas) onto fields. Although arable lands in La Cañada are limited to alluvial fans where tributaries enter the Río de las Vueltas or the Río Grande, population pressure did not demand more intensive hydraulic constructions, nor were these readily feasible.[49] Water diversion and transportation structures appeared only after the Zapotec conquest. We can imagine Zapotec hydraulic engineers sneering at the primitive state of La Cañada irrigation at the time of their conquest and ordering the vanquished serfs to start digging.

Between 300 and 200 BCE, a time frame that marks the end of the Perdido phase and the beginning of the Lomas phase (which corresponds roughly with the Monte Albán Late I) Zapotecs of Monte Albán conquered the region, killed off many of its inhabitants, completely destroyed existing settlements, eradicated the Indigenous culture, forced the abandonment of settlement sites, and relocated the people and villages to more defensible sites on terraces or hillsides above the farmlands in the canyon bottom. The Llano Perdido near present-day Santiago Dominguillo was abandoned and the populace forcibly relocated to the sides of the nearby hill known locally as Loma de la Coyotera. Residences became individual (rather than communal or multifamily), in the style of the Zapotecs, constructed on terraces carved into the hillside at what must have been considerable labor expense. In addition, Zapotecs appear to have ordered the construction of irrigation canals as a means for increasing agricultural production for commodities to be sent to Monte Albán as tribute or perhaps somewhat later to support the staff of the fortress at Quiotepec.[50] They remained in the Cañada through the Lomas phase, until roughly 200 CE, an expanse of half a millennium.

The emphatic nature of the Zapotec conquest is represented by a grisly *tzompantli* (skull rack) unearthed at Loma de la Coyotera, the settlement carved into a hillside above the Llano Perdido. Spencer and Redmond and their team unearthed sixty-one individual skulls plus numerous skull fragments at the site. When the skulls were excavated, they lay positioned in ordered rows, suggesting they had been mounted on a rack similar to those Aztecs used for public display following the defeat of an enemy.

49. Spencer and Redmond 1997, 599.

50. Spencer and Redmond 1997, 530.

The Loma de la Coyotera rack was assembled more than a millennium and a half before the Aztecs incorporated the skull rack into their symbolism. It represented a stark billboard to the conquered peoples of La Cañada, reminding them of the Zapotecs' intent to rule the region without any hint of resistance from residents. Spencer and Redmond suggest that skulls may have been added over time, since sixty-one skulls would have represented a large proportion of the existing population and a needless destruction of a much-needed labor force. Instead, skulls of rebellious or uncooperative individuals who needed to be punished appear to have been added as a reminder of Zapotec supremacy to the populace as a whole.

Construction of the fortress on the Cerro de Quiotepec appears to have followed the destruction of Llano Perdido, but considerable time may have elapsed between the two. Spencer and Redmond suggest that the new structures at Quiotepec may have been a Zapotec response to a perceived threat from the quickly developing financial and trade empire of Teotihuacan. Whatever the reasons, the infrastructure on the summit of the cerro and associated structures in the vicinity indicate an urgent response by the lords of Monte Albán to a threat from somewhere. The elaborate complex, which includes ceremonial structures as well as purely military ones, demonstrates that the Zapotecs intended to stay for the long haul. That they halted the importation of goods from the north is demonstrated by the preponderance of Lomas phase pottery, which is like pottery from Monte Albán of the same dates, and by the corresponding absence of pottery from Tehuacán or other sources to the north during the Zapotec occupation.

Sometime after 200 CE, a new era, the Trujano phase, appears in La Cañada. The acreage dedicated to agriculture shrank, as did the quantity of Zapotec artifacts, suggesting that the importance of the valley to Zapotecs waned or that their ability to provide the requisite administration sagged. It may also represent a military conquest of Monte Albán by Teotihuacan in the fourth century CE, as Winter suggests.[51] Although population in the vicinity of the Cerro Quiotepec increased, the fortress itself saw no further development and slowly fell into ruin or at least into disuse as a military installation. The population increase was probably a response by local inhabitants to attacks by various peoples and the absence of a central military authority to protect them. The easiest place in the entirety of La Cañada to fend off attackers or raiders would still have been the Cerro Quiotepec. In other words, La Cañada became a location of frequently contested ground, and the once-mighty fortress retained its defensive quality, but on a local, not an imperial basis.

The Quiotepec fortress is not the sole example of monumental architecture in the region of the confluence of the valleys. Nineteenth-century reports noted a site near

51. Winter 2007, 34.

Tecomavaca, some 10 kilometers north, containing many pyramids, stone monoliths, and burial sites all in good condition.[52] The dates of occupation and ethnicity of the peoples involved are not known. Adolfo Bandelier remarked in 1884 that Tecomavaca had been a Mazatec town, though the language was fast disappearing. Much of the area has now been converted to citrus orchards and milpas. Similar structures occupied a site nearby at Los Cues, and another at Teotitlán de Flores Magón.

The withdrawal of Zapotec forces from La Cañada around the end of the Lomas phase, that is, during the century following 200 CE, requires some discussion. Monte Albán would continue to expand and consolidate its rule over much of Oaxaca until well after the seventh century CE, more than four hundred years after its withdrawal or the phasing out of the Cerro Quiotepec. In other words, the abandonment of La Cañada was not correlated with a diminishing Oaxacan influence. Monte Albán's cultural importance continued to expand well after the departure of troops from La Cañada.

No evidence exists to suggest that the Zapotecs withdrew under military pressure from outside the region, though Winter detected dominance from Teotihuacan. Nor is there any ecological evidence that famine or other hardships forced the retreat. The withdrawal may have been an early sign of the weakening of Monte Albán's grip on its empire, but the most plausible explanation is that the Zapotec kings no longer viewed the fortress as necessary. The easing of military presence in La Cañada coincided with the rise of Teotihuacan and the founding of a Zapotec barrio in that distant megapolis. Trade increased dramatically across the valleys as Olmec, Maya, and Teotihuacan merchants expanded their movement of goods through the region. The Zapotecs could not have maintained a vigorous embargo on foreign goods and travelers while themselves promoting their empire and its goods in a foreign capital. The flow of goods and services may simply have overwhelmed the ability of the fortress to control the border, or the kings may have decided that the expense of maintaining the remote facility was simply not worth the effort. Trade, in other words, opened the border. Thin Orange pottery, a prominent symbol of Teotihuacan trade, appeared for the first time at some locations in La Cañada.[53] Other peoples occupied the Cerro Quiotepec infrastructure, but never again would it function as a military outpost controlling trade in the region.

La Cañada and its environs undoubtedly changed hands over the centuries following the retreat of the Zapotecs. Late in Postclassic times, several Cuicatec *cacicazgos* came to control the valley, none of them exceptionally powerful and none of them actively at war with any of the others. Under the rule of King Moctezuma II

52. Bandelier 1884, 267; Spencer and Redmond 1997, 627.

53. Spencer and Redmond 1997, 609.

early in the sixteenth century, Aztecs conquered the Cuicatecs and forced the various *cacicazgos* to pay tribute to Tenochtitlán via the Aztec administration center in Coixtlahuaca.[54] None of the several *cacicazgos* was able to challenge Aztec aggression, and submission came quickly and apparently without resistance. The Aztec reign did not last long. Spanish conquerors assumed the power position of the Aztecs by the 1530s and merely supplanted them as parasites of the colonies, extracting tribute through the encomienda, but not demanding captives to be executed in Mexico City.[55] The labor of serfs enriched the lords more than their executions.

At the time of the Conquest, the towns of Quiotepec and Tecomavaca were apparently under Mazatec control, while Cuicatlán was Cuicatec, as it had been for several centuries. Two centuries later Quiotepec was home to Cuicatecs and Mazatecs. In 1884, Bandelier found Quiotepec to be a Mazatec town with few Mazatec speakers.[56] Today, although Cuicatec and Mixtec speakers are usually to be found in the market in Cuicatlán, these are visitors from mountainside towns to the east and west. Most residents of Cuicatlán speak only Spanish. Cuicatecs have long since been dislodged by Hispanic caciques and hacendados and relegated to the steep sides of the Sierra Juárez, while the Mixtecs visit the city only to market their wares. Their homes are in lofty lands to the west, along with the homelands of a smattering of Chochos and the few remaining Ixcatecs. The residents of Quiotepec are aware that the Cerro Quiotepec is ancient and once housed Zapotecs, but they view the site as primarily a source of employment.

We have visited the site several times. In addition to its commanding views and impressive archaeological remnants, the site presents an intriguing botanical heritage. In earlier trips we noted the presence of unusually large numbers of agaves and columnar cacti among the ruins. In early 2016 we were joined by agave experts Daniel Serrano and Greg Starr, who joined us after seeing our photographs of the site. An examination of the cerro revealed a proliferation of agaves of unclear taxonomy, some of which appear not to be found elsewhere.

After considerable discussion we offer the tantalizing possibility that over the centuries, the residents of Cerro Quiotepec transplanted to the summit several agaves and cacti (and perhaps other plants) that were exceptionally useful to them. Agaves provide a source of food, fiber, and medicine, while the two most prominent columnar cacti (*Escontria chiotilla* and *Pachycereus weberi*) are both producers of abundant crops of fruit, the latter a source of wine and shade as well. Since delivery

54. Hunt 1972, 209.

55. The encomienda was a grant by the Spanish sovereign to a worthy subject. It consisted of a tract of land with natives that inhabited it. The encomendero was authorized (with some restrictions) to tax the residents as he saw fit but could not reside on the land.

56. Spencer and Redmond 1997, 620–21; Hunt 1972, 167; Bandelier 1884, 266–68.

of food and fiber to the summit was laborious and time consuming, the presence of useful plants there would be a nutritional and technical convenience. The unexpected presence of agaves unusual to the area, the isolation of the agaves from populations of similar species, and the many centuries available for emergence of new taxa suggest remarkable possibilities of human intervention in plant evolution.

Tehuacán Viejo (La Mesa)

Ninety kilometers northwest of Cerro Quiotepec and 1,000 meters higher in elevation lies one of Mexico's newly celebrated archaeological treasures, Tehuacán Viejo, also popularly known as El Viejo and La Mesa. This spectacular 125 hectare (309 acre) site, located only about 8 kilometers from Tehuacán's central plaza, is situated atop a mesa that ramps upward toward the east. Looming above the ruins rise the ramparts of an escarpment known locally as Cerro Colorado, an outcropping of the Sierra Zongolica (and the Oaxaca Fault) above the eastern limits of the Valley of Tehuacán. From the summit of the mesa, Popolocan authorities could have commanded a broad view of the valley and its activities, while the monumental buildings and plumes of incense smoke would also have been visible for many kilometers from below. As of this writing only 10 percent of the site has undergone excavation. The site is now an official INAH site, complete with a museum and guides. When we first visited La Mesa, it languished in magnificent solitude, patrolled by a solitary guard, an employee of Tehuacán's municipal government. In the early 1990s INAH conducted preliminary reconnaissance and initial excavation, but funding for full-blown excavation and interpretation barely trickled down. The ruins were usually visited for a few weeks each year by a small team of Mexican archaeologists, who heroically maintained the restoration and site reconnaissance work and struggled against the encroachment of lush thornscrub vegetation merging into tropical deciduous forest.

INAH has officially labeled the site Tehuacán el Viejo, also referring to it as La Mesa. It is fenced, protected by chain link from looters, woodcutters, and pastoralists, all to the great advantage of the surrounding forest. Pronouncing an official name for the site was controversial, since some writers insist that Tehuacán Viejo refers to a site at the base of the mesa where Franciscan monks constructed a monastery early in the sixteenth century. La Mesa, others argue, refers simply to the large mesa on which the site was constructed. Its closeness to bustling Tehuacán has elevated Tehuacán el Viejo (La Mesa) to one of the most visited archaeological sites in Mexico.[57]

57. Yomara Pacheco, "'La Mesa' entre las 16 más visitadas en Tehuacán," *El Popular*, June 29, 2018, https://elpopular.mx/secciones/municipios/2018/06/29/la-mesa-entre-las-16-mas-visitadas-en-tehuacan.

FIGURE 6.21 Civic and ceremonial structure dating from 750–900 CE, La Mesa, Tehuacán. In the background rise the cliffs of the Oaxaca Fault.

It is as it should be.

Tehuacán el Viejo's archaeological importance has been recognized since the beginning of the twentieth century.[58] Joaquín Paredes Colín, noted Tehuacán historian, makes only passing references to the site of La Mesa in his history of the city and region, written in 1910. Richard MacNeish, while familiar with Cuthá, was apparently unaware of the size and sophistication of the Tehuacán Viejo site, for he makes no mention of it. He wrote in 1967:

> The [Tehuacán] valley was probably divided into city-states or kingdoms made up of towns and hamlets surrounding urban centers, often fortified, situated in the flanks of the valley in the same locations as present-day towns. Camps, shrines, villages, salt making sites, and quarry-sites, guardhouses and habitations connect with a wide range of irrigation features were spread throughout the region. Economy was based on irrigation agriculture, commerce, salt-production, processing of cotton, and making various stone implements. . . . Social organization was intricate and stratified.[59]

Until 1993 the ruined buildings of La Mesa remained covered with dense tropical deciduous forest and resembled amorphous mounds. At the time of INAH's initial

58. Jäecklein 1979, 203.

59. MacNeish 1967, 24. This should not be interpreted as a criticism. Archaeology of the Classic and Postclassic sites was beyond the scope of the Tehuacán Project, which dealt primarily with origins of agriculture in the Valley of Tehuacán and its development through time.

excavations in 1993, the site was included in the communally owned lands of an ejido.[60] Ejidatarios used the lands to graze their animals, mostly goats, and gather firewood. Negotiations for transferring the lands to the jurisdiction of INAH were successful, and roughly 200 hectares of land is now controlled by the archaeological bureau of the Mexican government.[61]

A consensus of investigators dates occupation of the Tehuacán Viejo site from the late Preclassic (500–150 BCE), but its largest monuments are of much later construction. Archaeologist Noemí Castillo Tejero provides the following description (translation ours):

> The archaeological site indicates a continuous occupation apparently from the end of the Classic Period [ca. 750–900 CE] with a growth from the north to the south, following the Mesoamerican system of orientation, that includes a series of linked plazas framed by various structures, among which stand out pyramid bases, hypostyle halls, low platforms on which were built portico rooms as well as residential buildings, the so-called palaces, although the peasant population would have lived outside La Mesa, in what up to now is the fertile Valle of Tehuacán.[62]

The site was occupied and then abandoned during the Late Classic and reoccupied in the Postclassic, during which time the native Popolocas constructed the monumental buildings. Construction continued until the conquest by Aztecs in the mid-fifteenth century.[63] Recent excavations have revealed a fourteenth-century shrine probably dedicated to Mictlantecuhtli, the Aztec Lord of Death, the only such shrine known. The structure lies buried beneath the plaza of the large partially reconstructed main pyramid complex. The building contains human skulls embedded in the structure's walls and ceramic images of a skull with red painted tongue thought to be representations of the Lord of Death that resemble images illustrated in the Codex Borgia.[64] Numerous human bone remnants also suggest that the structure was a site for human sacrifice. INAH archaeologists have named the building the Temple of Skulls.

The excavated portion of Tehuacán Viejo consists of three plazas and a total of twenty structures. The largest of these are monumental structures with pyramid

60. An ejido is a system of land tenure that originated in Spain. Since the land reform policies of the Mexican Revolution, ejidos have become cooperatively owned tracts of land usually with individually assigned parcels. *Comunidades* are tracts of communally owned lands, the members usually with pre-Hispanic origins.

61. Castillo Tejero 2000, 303.

62. Castillo Tejero 2002, 73.

63. Gámez 2003, 70.

64. Códice Borgia 1976.

FIGURE 6.22 La Mesa, Valley of Tehuacán, archaeological site with Cerro Colorado, a portion of the Oaxaca Fault. Note the rich tropical deciduous forest.

bases finished with temples and altars at the top of steep staircases. Two structures are semicircular, connecting in the rear to rectangular buildings. The largest plaza contains a pyramid that rises more than 15 meters above the leveled plaza, affording a grand view from above and an impressive elevation from below. Altars are commonplace throughout the site, demonstrating its religious importance. Burials indicate an added ceremonial dimension, and human bones throughout suggest that human sacrifice was routinely practiced. One habitational structure is of such size and elegance that the researchers consider it to have been a palace. The amount of construction and importation of materials for the buildings on La Mesa would have required a host of planners and engineers in addition to large crews of laborers of varying degrees of skill to complete the work on the site. The architects were artists and artisans of the first rank.

This temporal classification tells us little about the nature of the builders and their culture, except that the site contains themes in common with Nahuas to the north. During Postclassic times residents of Tehuacán would have experienced ongoing relations—friendly, hostile, or neutral—with Chochos, Cholulans (probably Olmeca-Xicalanca), Mazatecs, Maya, Mixtecs, a variety of Nahua-speaking peoples, and Totonacans (from western Veracruz). Some authorities suggest that the site may have been overrun by Toltecs after the collapse of Tula in about 1200 CE, following their conquest of Cholula. Castillo Tejero believes that Tehuacán Viejo was the site of one of four Popolocan *señoríos*, roughly equivalent to feudal estates.[65] Oral histories and codices place the date of Aztec conquest of the area at somewhere between 1456 and 1461 CE, during the reign of Moctezuma I, at which time construction on part of the temple complex ceased, and the Aztecs forced the Popolocas to relocate the seat of their *señorío* to the base of the mesa at a site called Calcahualco, again an Aztec name.[66] The name of their kingdom thus became Tehuacán, and Popolocas became tribute payers to Tenochtitlán. After the Aztec conquest, Moctezuma's forces battled their way south as far as Tehuantepec, probably relying on the productivity of the Valley of Tehuacán to replenish their supply lines between Tehuacán and the rich fields of the central valleys of Oaxaca.

Franciscan friars believed that Popolocas demonstrated an affinity for Christianity. The Brown Robes thus undertook construction of the first site of Tehuacán in 1540 at the base of La Mesa in the hopes that by immersion within the population center, they would bring a quick conversion of the natives. That effort was a dismal failure. The site seems to have been cursed with ongoing plagues of disease, an

65. Castillo Tejero ca. 2008. The other three *señoríos* were Tepexi and Tecamachalco in Puebla and Coixtlahuaca in Oaxaca.

66. "Shrine Dedicated to God of Death Unearthed in Mexico," *Archaeology*, November 21, 2013. https://www.archaeology.org/news/1546–131121-mexico-temple-skulls.

FIGURE 6.23 La Mesa. Part of the large central plaza. The mound with the large trees is a small pyramidal temple. Many buildings have yet to be excavated.

alarming abundance of rattlesnakes, and relentless attacks by biting ants.[67] Admitting defeat, in 1568 the Franciscans relocated their mission to the present site of the convent in the city of Tehuacán. Tehuacán Viejo and its rich archaeological past receded into obscurity. Only recently has it welcomed visitors. They can now appreciate the sophistication of Popolocan architects and engineers and their wisdom in selecting a site rich in beauty.

67. Paredes Colín (1910) 1953, 59–60.

CHAPTER 7

Plants and Ecosystems

The valleys of Cuicatlán and Tehuacán host a wealth of archaeological, cultural, historical, and geological features, sufficient to make them extraordinary. But it is the plants that set the valleys apart from anywhere else in the world. To the trained ecologist and the plant enthusiast, the valleys present almost unending examples of endemism, the ecological condition in which a species is found only in a restricted location. In addition, the way the plants comingle, forming what ecologists call plant communities or associations, are so varied, so distinct, and so unusual that they alone would justify naming the region as a Biosphere Reserve. Together, these unique plants and their vegetation associates made these valleys world famous long before systematic archaeological work was undertaken and before ethnohistorical studies revealed the remarkable convergence of cultures in the valleys. Jerzy Rzedowski, the dean of Mexican field botanists, states that the upper Papaloapan basin, which includes Cuicatlán and Tehuacán Valleys, is the most ecologically diverse region of Mexico.[1] And Mexico houses one of the world's highest rates of biodiversity.

The plant communities—and they are many—range from forests of tropical deciduous trees rich in *Burseras* and legumes to high-elevation pines and firs, from vast oak woodlands to thick chaparral, from dense stands of desert columnar cacti to thick groves of tall palms, and a host of other associations found in between, including rich agroecosystems, where cultivated plants have coevolved for millennia with weedy plants growing in habitats disturbed by humans. While it is tempting to describe these associations as discrete vegetation units, a closer examination reveals a remarkable continuity from the dry plant regimes in the valley bottoms to

1. Rzedowski, Medina-Lémus, and Calderón de Rzedowski 2004, 28.

FIGURE 7.1 Mixed desert and thornscrub landscape, Calapa Canyon, Valley of Tehuacán.

the humid montane forests of the high sierras. Tropical and boreal elements and vegetation of both North American and South American origin interpenetrate at all levels: plants from the Neotropics coexist with conifers of Nearctic origin at the higher elevations, where palms and pines freely share the landscape. In the valley bottoms we find a flora that is a complex mixture of species whose evolution is traced to South America (cacti, mesquites, acacias, mimosas, and palo verdes), quasi-endemic Mexican plant groups (Agavaceae, Fouquieriaceae, and North American columnar cacti), and other new groups. This proliferation of plant associations resulted from the explosive diversification of dry-land plant groups that appeared throughout the world during the global drying and cooling trend that began in the Miocene (23–5 MYA).[2] Oaks, of which Mexico presents an astounding variety, are important components of several plant communities, ranging from minor shrubs to majestic, dominant trees. The unusual climatic conditions found in the valleys, combined with their complex geology, have made them an evolutionary laboratory.

This tapestry of intermingled species and vegetation types represents the true botanical richness of the valleys. The charismatic plants—agaves, barrel cacti, columnar cacti, Beaucarneas, Burseras, cycads, Fouquierias (the ocotillo family), oaks,

2. Arakaki et al. 2011.

palms, and yuccas—and even less conspicuous endemics such as ground-hugging cacti and delicate succulents combine in ways that provide a seemingly endless variety of plant landscapes. Endemic species range from the giant columnar cactus *Neobuxbaumia macrocephala* to the diminutive succulent *Echeveria laui*. A recent survey notes 365 endemic plant species, or about 14 percent of the known flora of the valleys, well over 2,600 species. More than 40 percent of the 180 plant families include at least one endemic species.[3] This highly localized diversity adds to Mexico's status as a hotspot of biodiversity, with 24,360 known vascular plant species of which at least 10,235 (42 percent) are endemic.[4]

Classifying the vegetation types—the variety of plant associations—found in the valleys presents a challenge. It is important to distinguish between vegetation types and plant communities. Vegetation types are more related to overall appearance, while plant communities stress ecological associations as well as floristic compositions. One can recognize at a glance the vegetation type of a landscape, but perceiving its ecological associations requires further analysis, depending on the degree of specificity required or the purposes of the describer. Less conspicuous shrubs and herbs may exert a widespread effect on ecological relationships.

The plant list of the region is vast. Recent estimates for the flora of the state of Oaxaca alone report 9,589 plant taxa (9,362 species and 227 infraspecific categories distributed among 321 families and 2,160 genera). Ongoing research will likely show that the flora of Oaxaca exceeds ten thousand species, a veritable hotspot of North American species diversity.[5] If we add southern Puebla, which includes the Valley of Tehuacán to the count, the numbers will be even greater. Given this immense diversity, one could easily describe twenty different types of vegetation and still be nervous about oversimplification. We present a mere handful of vegetation types and acknowledge that one can readily discern more, perhaps many more. We caution the reader that such classifications reflect a combination of steepness and aspect of the slopes, soil type, elevation, geology, orientation, and geographic location (in reference to rain shadow effects of mountain ranges, for example). Step a few paces from a northern slope to a western exposure and the change in plant composition may be startling. An alteration in any of the variables can produce different vegetation in a short space, so these descriptions are only to be applied generally. We have made roughly twenty trips through the valleys. Each visit brings us new observations on

3. Dávila et al. 2002, 421. By comparison, the Yunnan Great Rivers area of southwest China is the richest in plant diversity in Asia, with 7,000 plant species, of which 13 percent are endemic. It is largely rain and cloud forest. See Chang-Le et al. 2007, 759. California has around 6,000 species and is five times as large as the state of Oaxaca.

4. Sosa, De-Nova, and Vázquez-Cruz 2018, 524.

5. García-Mendoza and Meave 2011.

the vegetation, often in conflict with what we previously believed. Our education is ongoing and lifelong. One day we stopped our vehicle in what appeared to be an arid, depauperate landscape. Within a few paces we looked in surprise at each other as we discovered a variety of plants far beyond what a glimpse at the vegetation suggested.

In 1965, as part of the Tehuacán Project, C. Earle Smith published the flora of the Tehuacán Valley, also describing the vegetation, basing his classification on earlier fieldwork that distinguished botanist Faustino Miranda published in 1948. We expand modestly on the vegetation classifications of Miranda and Smith while acknowledging their perspicuity. We note also that their studies were confined to the Valley of Tehuacán and did not include La Cañada, which adds a minor host of vegetation types and plant species to the list.

The works of Miranda and of Smith were by no means the first attention paid to the extraordinary diversity of the valleys. Well before the arrival of Europeans, natural history and live botanical collections in gardens were prized among the ancient Native American cultures. Pre-Hispanic gardens were the exclusive domain of the Mesoamerican upper class. As was the case in Europe at the time of Conquest, formal gardens were a sign of prestige, power, and grandeur. According to sixteenth-century Spanish historian Francisco López de Gómara, the Aztec king Moctezuma

> had wondrous gardens with medicinal and fragrant herbs, with flowers, with roses, with perfumed trees, that are infinite. It was to praise the creator so much diversity, so much freshness and scents. . . . Moctezuma did not consent to have either fruits nor vegetables in this oasis saying that was not appropriate of kings to have such lowly enterprises in these delightful places; those orchards were for slaves or merchants. . . . He also had fountains, rivers, pools with fish, rabbit hutches, pens, cliffs, and ridges where deer, stags, hares, foxes, wolves and other animals [roamed].[6]

Zoos and gardens were probably kept, maintained, and enjoyed in all powerful Mesoamerican cultures. They served not only as expressions of power and leisure, but also as a ready supply of important medicinal resources, for acclimatizing and domesticating new crops and animals, and for providing the court with exotic fruits and animals for consumption. Since the 1400s central Mexican rulers had enjoyed splendid parks and collections of live animals and plants. For example, Netzahualcoyotl, the philosopher-king, had leisure gardens in his palace in Texcoco, horticultural gardens covering more than 200 acres in Texcotzingo, and game reserves with exotic plants and animals elsewhere in his kingdom.[7] The same customs undoubt-

6. López de Gómara (1552) 1979, 121.

7. Evans 2010.

edly prevailed in older cultures, like those of Teotihuacan, northeast of modern-day Mexico City; the Mixtec-Zapotec kingdoms in the lofty valleys of Oaxaca; or the Popolocan statelet reigning over the salinas at Cuthá in the Valley of Zapotitlán as they re-created nature within their urban centers.[8]

Early conquistadores expressed their wonder at pre-Columbian zoos. Hernán Cortés, in his 1520 *segunda carta de relación* to King Carlos I of the Spain, described in detail the "animal houses"—*totocalli*—which harbored extensive animal collections.[9] Later, in his third letter to the king, he deplored the destruction of Moctezuma's zoo.[10] Recent research in the Templo Mayor in Mexico City produced remains of more than sixty vertebrate species, from scorpion fishes to jaguars, including twenty-eight fish, eight reptile, twelve bird, and twelve mammal species.[11] These animal houses not only were present in the vanquished city, but were undoubtedly a normal occurrence throughout Mesoamerica and its area of influence. Findings of captive populations of different tropical macaw species as far north as the Pueblo cultures of northern Mexico and the southwestern United States, of captive wolves in Chiapas near the border with Guatemala, and of bighorn sheep remains in locations north of Mexico City attest to the prevalence of these types of collections.[12]

Modern botanical collections in the valleys started formally with the work of Cyrus G. Pringle, one of the most prolific plant collectors in history. He made thirty-nine expeditions to Mexico between 1885 and 1909, and collected more than half a million specimens, many of them species new to science.[13] His last expedition in 1909 was made against the advice of his field assistant, Filemón Lozano, who warned of the dangers of unrest in the impending Mexican Revolution of 1910. Pringle collected throughout Mexico, including La Cañada region and the Valley of Tehuacán. He boarded the train in Charlotte, Vermont, for his southbound explorations, an appropriate departure location given the absence of major roads in Mexico. He journeyed on the train to Mexico City, then by more rustic transportation south to Tehuacán and beyond toward Oaxaca.

Pringle sometimes traveled with other famous botanists, among whom were Edward Palmer, the father of ethnobotany; E. W. Nelson, the most important authority on Mexican birds at the time and a remarkable ecologist; and Major E. A. Goldman, well known for his vertebrate descriptions and ecological work. While

8. Morales Folguera 2004, 352ff.

9. King Carlos I of Spain was also known as King Charles V, the Holy Roman Emperor famous for the phrase "The empire on which the sun never sets."

10. Cortés 1963.

11. López 1993; Blanco et al. 2009, 28–39.

12. Whalen and Minnis 2009, 242–46; Manin and Lefèvre 2016, 135.

13. Nicholson 2002, 2–13.

FIGURE 7.2 Cyrus G. Pringle, early plant explorer in the valleys.

working for the Smithsonian Institution, Pringle stimulated the ongoing work of Cassiano Conzatti, an Italian botanist, with whom he journeyed to Tomellín Canyon in La Cañada in 1897. Conzatti and Pringle collaborated and had a long friendship. Conzatti continued his work on botanical endeavors until 1937, mainly in the region of La Cañada. Many of his collections are now housed at the National Herbarium of Mexico and the Smithsonian Institution, but the bulk of his specimens disappeared as a result of neglect and lack of interest. Conzatti's prominence is commemorated in a shaded plaza that bears his name in the city of Oaxaca.

In 1907, German plant collector Karl Purpus began gathering plants in the valleys. In 1912, during the Revolución Mexicana, he stated, "Because of the conditions of insecurity that prevail in many states, I could only explore in Veracruz and part of Puebla. Thus, I visited Tehuacán and Esperanza. It would have been very risky to make more distant excursions. The *Zapatistas* were everywhere."[14] He gathered specimens with famous botanist Townshend Brandegee, who helped him in the botanical collecting but whose name never appears in Purpus's journals.[15] At roughly the same time, J. S. Rose, co-author with N. L. Britton of a classic and influential publication, *The Cactaceae*, collected in the valleys as well.

The tradition of Mexican collectors and taxonomists was well established by the eighteenth century, and extraordinary Mexican biologists were influential and productive during the nineteen century.[16] The earliest recorded formal biological exploration in the valleys, however, was not undertaken until the 1940s, when Faustino

14. Purpus appears to have been a victim of the propaganda machines of Porfirio Diaz. The Zapatistas were powerful revolutionaries but most careful about selecting their targets—the dictatorial regime and its wealthy national and international enablers. No record of harassment of field scientists, domestic or foreign, by Zapatistas has emerged from the documents of the Mexican Revolution.

15. Sousa 1969, 1–36. Sousa cites Otto Nagel, a respected orchidologist, as stating that Purpus was "a difficult man, almost puritanical. He did not take alcohol in any form; he did not smoke and apparently because of an unhappy love in his youth he was very reserved in his feminine relationships."

16. Eighteenth-century Mexican botanists include, among others, José Antonio Alzate y Ramírez, Francisco Javier Clavijero, José Mariano Moçiño Suárez Lozada, Martín Sessé y Lacasta, and Longinos Martínez. For details, see Michan Aguirre and Llorente Bousquets 2003. Nineteenth-century biologists include Alfredo

FIGURE 7.3 Cassiano Conzatti, late nineteenth- and early twentieth-century botanist. He has a plaza named after him in the city of Oaxaca.

FIGURE 7.4 Karl Purpus, eccentric but accomplished botanist of the valleys.

Miranda, a Spanish botanist, began his field research. He had migrated to Mexico along with many other Spanish intellectuals after the collapse of the Spanish Republic in 1939. Miranda was the first to characterize the vegetation of the valleys, proposing eight vegetation types in the La Cañada region.[17] His work stimulated a botanical and ecological revolution in Mexico that produced the first modern treatise on its vegetation.[18] It was later formalized by Rzedowski.

Since publication of the pioneering work by Miranda and the synthesis by Rzedowski, dozens of researchers and their students have documented the rich flora and vegetation of the valleys. Others have described biotic interactions found only in the valleys, a work well researched by Reyes Santiago et al. for the region of La Cañada.[19] Among the scholars researching the plants of the valleys, Fernando Chiang Cabrera, Francisco González Medrano, and many of their students merit special mention for their role in producing and publishing information about the flora of the valleys.

Augusto Dugés, Alfonso Herrera Fernández, José Ramírez, José María Velasco, Francisco del Paso y Troncoso, and Alfonso Luis Herrera, among others.

17. Miranda 1942, 417–59; 1948, 333–64; Pérez-Malváez et al. 2007.

18. Miranda and Hernández-Xolocotzi 1963, 29–179.

19. Reyes Santiago et al. 2004.

Animal Life

Although the entire region was rich in animal species during the Late Pleistocene, animal life in these valleys and uplands is now sparse by comparison with plant life. Wild mammals are notably lacking. The long history of human occupation—at least ten thousand years and well populated for at least the last two millennia—with the inevitable consumption of resources and persecution of huntable animals has been hard on the fauna. Most predators, major herbivores, and all snakes are still relentlessly persecuted. Unending and unrestricted grazing by European-introduced herbivores has inflicted five centuries of stress on the vegetation—stamping, nibbling, chewing, tearing, scratching, despoiling the land, while disturbing and compacting the soil, favoring the proliferation of weedy and tough plants that are unpalatable or toxic to livestock and wildlife. Habitats where wildlife can find refuge are limited and shrinking. As a result, most charismatic wildlife has been exterminated or has retreated to the most isolated recesses of the mountains. Apart from occasional coyotes, foxes, and perhaps a stray bobcat, ocelot, or jaguarundi in the remote mountains, one seldom sees wildlife larger than rodents, jackrabbits, or opossums.[20] Deer are occasionally mentioned by locals, though in all our travels throughout the valleys, we have yet to spy one, so we must assume they are rare. An informant who took us to an isolated mountain site where we found the cycad *Dioon purpusii* mentioned that his goats had been attacked by a "tigre," the local name for a jaguar.

Bat species, on the other hand, continue to find the valleys much to their liking. Researchers have documented thirty-four species of bats in the Valley of Tehuacán, and additional species will be added with further investigations in the Valley of Cuicatlán. Bats are intimately involved with the proliferation and distribution of columnar cacti. Many species of columnars are pollinated exclusively or primarily by bats.[21] A cave near Zinacatepec ("near the hill of bats" in Náhuatl: *tzanacan* = bat; *tépetl* = hill), protected by a chapel located on the hill above, provides an excellent roosting site for a colony of bats that are critical to pollination of huge numbers of cacti in the Valley of Tehuacán.

---

20. Búrquez spotted what appeared to be a jaguarundi in the eastern slopes of the Sierra Zapotitlán in 2013. Residents of Zapotitlán have warned us that a small lizard that frequents agaves and yuccas is highly venomous. It is not. In our many visits to the monte of the valleys, we have yet to encounter a single wild venomous snake and have spotted only three snakes, two of them harmless, the third a rattlesnake inside a protected area, mortally wounded by a human attack. On the eastern slopes of the Sierra Juárez, the steepness of the upper slopes (as opposed to the foothills) and the density of vegetation have discouraged clearing of the landscapes for pasture or milpas. Pathways through the forest, a wet, mesophilic rainforest, are few, but kinkajous, coatimundis, and small cats such as ocelots and jaguarundis, as well as several species of monkeys, are still occasionally seen.

21. Dávila et al. 2002, 436.

FIGURE 7.5 Cave near Zinacatepec that shelters a large colony of bats.

In comparison with mammals, birds in both species number and individuals are abundant, but apart from a few charismatic species, their distribution, nesting, and occurrence are not well documented. Various sources list between 330 and 360 bird species that have been observed in the valleys, but this number is surely no more than a fraction of the number that will emerge from a more rigorous bird survey. Many of those observed are migratory birds that spend wintertime in the mild climate of the valleys and await springtime to fly north to their breeding grounds. Birds and bats, highly mobile organisms, probably share resources better than terrestrial mammals because they can exploit resources derived from various regional ecosystems across the elevational gradient.

Notable among resident birds is the sanctuary for a sizable population of the threatened military macaws that nest in the steep walls of the deep Cañón del Sabino, located in the southwesternmost reaches of the Valley of Tehuacán. The canyon is a well-protected reserve and lies under the jurisdiction of the Oaxacan community of Tecomavaca.

Insects are plentiful throughout the year. Since antiquity, many of them have been incorporated into the human menu, finding their way into the native cuisine by the special qualities of their flavor, for their nutritional value, and because of the scarcity of other forms of protein. Insects are not emergency food, however, but

FIGURE 7.6 *Cuchamá* (larva of saturniid moth, *Paradirphia fumosa*) and cactus fruit juice, *agua de garambullo*, Zapotitlán Salinas.

constitute special and luxurious delicacies that complemented and added zest and flavor to the corn-bean-squash diet of the ancient Mexicans. Bernardino de Sahagún in the Florentine Codex listed ninety-six species of edible insects among the Aztecs. Julieta Ramos-Elorduy and colleagues have documented more than ninety species of edible insects still consumed in the central Mexican highlands and more than 535 useful insect species throughout central and southern Mexico.[22] In the valleys, the town of Reyes Metzontla has the distinction of featuring seventeen species of edible insects commonly served on the tables of the Popolocan inhabitants. These include *cuchamá*, the Popolocan name for the larva of a large saturniid moth (*Paradirphia fumosa*); several larvae of beetles that feast on agaves and other tropical plants, referred to collectively as *gusanos* (*gusano rojo* and *blanco del maguey*, *gusano del cazahuate*, *gusano del pirul*); and several species of bees and wasps whose larvae and honey are avidly sought after. Popular foods also include *chicatanas*, adult queens of the leaf-cutting ants that emerge after the first summer rains. These are ground to make a famous salsa. *Jumiles*, species of bugs that add strongly perfumed scents to food delicacies, are sometimes eaten alive.[23]

A Brief North-to-South Overview of the Vegetation Landscapes

Even with the variety of habitats, a keen observer can travel by highway from the lofty Central Plateau of Mexico, around the city of Cholula, down the ramp southeast

22. Ramos-Elorduy and Pino-Moreno 1989, 2005.

23. Acuña et al. 2011.

FIGURE 7.7 Fields near Zinacatepec, Valley of Tehuacán.

to Tehuacán without being struck by the celebrated diversity of plants. One can even continue for 10 kilometers or so south toward Oaxaca, either by the modern freeway, 135D or by the old Pan-American Highway, Mexico Route 135, without encountering anything impressive in the plant landscape. Most of the land northwest and southeast of Tehuacán is a seemingly unending mosaic of milpas—cornfields—thousands, tens of thousands of them, mostly small, hardly more than a hectare or two in size, of every possible shape, rectangular, square, trapezoidal, and irregular. It is as though every farmable square meter of the valley bottom is either urbanized or planted in pasture, milpa, or sugarcane, and almost no original vegetation remains to be seen. This is not a recent phenomenon, for this landscape was probably already a common agrarian scene hundreds of years ago, when countless ancient settlers farmed the extensive valley. Water and rich farmland were plentiful at the time. Today, the water supply for agriculture is imperiled by the burgeoning growth and thirst of the city of Tehuacán and a recent history of pumping groundwater for irrigating sugarcane.

Soon, though, as the ramp continues southeast, arable parcels of land dwindle and the arid sides of the valley, pediments, gentle slopes or steeper hills, close in. The highway now crosses long stretches of bajadas (coalescing alluvial fans along the base of the mountains). Below, the irrigated fields retreat as the narrowing valley limits land suitable for farming. On both sides of the road, the reality of decidedly peculiar plants becomes undeniable. Strange cacti, yucca-like sentinels, and several

FIGURE 7.8 Satellite view of fields near San Gabriel Chilac, Valley of Tehuacán. Google Earth.

species of gnarled trees of the genus *Bursera* (often called elephant trees)—short trees and bonsai-like shrubs almost unknown in the high, cool Basin of Mexico—close in on the highway. Thick groves of tall cacti stretch off into the distance. *Tetechos* (*Neobuxbaumia tetetzo*), garambullos (*Myrtillocactus geometrizans*), *xoconochtlis* (*Stenocereus stellatus*), and *viejitos* (*Cephalocereus columna-trajani*) loom along the roadsides or on the hillsides, whether as scattered individuals or in dense stands.[24] As the highway gradually descends toward the lower end of the Valley of Tehuacán, the massive candelabra cacti generically called *cardones* (*Pachycereus weberi*) appear here and there.

The highways, new and old, head generally south. They reach the low point in the valleys and, splitting ever farther apart, begin the long climb into the central valleys of Oaxaca. They gradually leave behind the cactus-dominated arid lands of thornscrub and rapidly ascend into tropical deciduous forests (TDFs). Though now much reduced, the TDF is still characterized by a multitude of tree species, most of which shed their leaves in the dry season. Several hundred meters above, the tropical deciduous forest zone merges into the belt of highly varied vegetation, much of it composed of dense shrubs and short trees that we will call *chaparral complex*. The chaparral is followed at higher elevation by the oak zone, its trees laden with orchids,

24. *Xoconochtlis* or *xoconostles*. In the written Náhuatl language, the letter *x* as the initial letter of a word usually has a sound like English *sh*: show-co-NOCH-tli. In most other cases, it sounds more like the German *ch*, as in Bach, or loch in Scottish. *Texas* was originally transcribed Te-shas and written Tejas.

bromeliads, and Spanish moss. Finally the road arrives at the oak-pine zone, which harbors its own cast of epiphytic plants, lichens, and mosses.

But the two roadways provide different introductions to the plant life of the valleys. The freeway, Mexico 135D, is blasted from steep hillsides, its cuts producing scars visible for tens of kilometers. It rises dramatically west of the valleys, first crossing the graceful bridge over the Río Calapa as it enters Oaxaca. It abruptly bypasses La Cañada and Cuicatlán and climbs across the Caltepec Fault zone. The wide highway gradually leaves below the dense groves of cactus-rich thornscrub, but not before revealing vast mountainsides of columnar cacti on the lower slopes of the Sierra Mixteca to the west and the sentinel-like mountain Cerro Verde to the east. It passes quickly through patches of dense tropical deciduous forest that remain somewhat protected by the precipitously steep slopes. After 10 kilometers of sharp ascent, the roadside vegetation gradually merges into chaparral and veritable forests of the goblet-shaped, many-branched cacti *jiotillas* (*Escontria chiotilla*). The sharp-eyed may catch a glimpse of the charismatic *jarilla* (*Fouquieria purpusii*) on a ridge above the expressway, though finding a space to park one's vehicle safely is an ongoing challenge.

Soon the desert vegetation is left far below, the chaparral thins, and the highway penetrates the Mixteca Alta, an ancient land once prosperous and productive but now impoverished and eroded into grotesque forms and dotted with towns in a state of slow abandonment. The remaining vegetation visible from the highway appears to be a compromise born of desperation between fragments of persistent, overgrazed plant communities of the past struggling for survival along with government-sponsored attempts at reforestation with pine trees—with varying degrees of success—and scraggly or abandoned milpas. Numerous ghostly terraces bespeak former agricultural prosperity, a time long ago when the region bustled with cornfields and villages. Some of the areas that appear to be overgrown with abandoned secondary derelict vegetation growing on lithosols (rocky soils) are surprisingly rich in species, with a dominance of distinct sclerophyllous chaparral elements.

Large areas of the ground here are eroded down to the substrate and support only low, tough shrubs, diminutive cacti, and resistant grasses, many of them with chaparral affiliations. The tough, whitish crust that has formed over the centuries on the surface of much of eastern Mixteca Alta and the Valley of Zapotitlán is variously referred to as *caliche*, *endeque*, or *tepetate*. Strictly speaking, caliche refers to a carbonate substrate that forms in arid soils as a result of limited soil penetration of rainfall and plant uptake of all available water. Using this characterization, the whitish substrate around Coixtlahuaca (rainfall over 500 mm annually) should not be considered caliche—although most likely some areas have caliche beds. Some sources refer to the hardened soils as *tepetate*—dense, indurated soil horizons, often

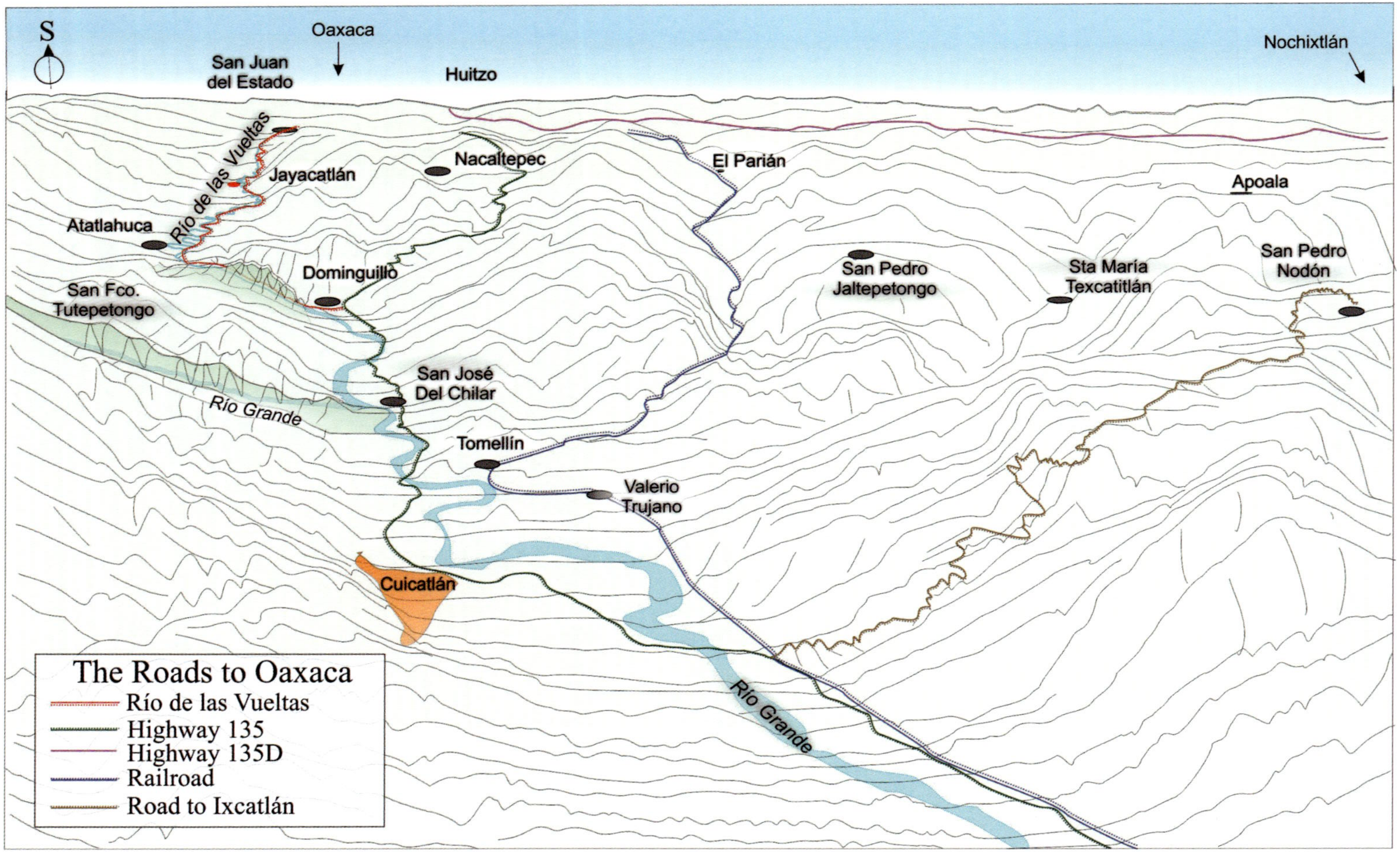

FIGURE 7.9 Diagram of the roads connecting Tehuacán and Oaxaca.

FIGURE 7.10 Calapa Bridge, Río Calapa (also called Río Hondo), looking northeast near the southern end of the Valley of Tehuacán. In the far distance beyond the mesa is the Sierra Zongolica.

FIGURE 7.11 Scars left from construction of Mexico Highway 135D, a toll road connecting Tehuacán and Oaxaca. The view is toward the south from Calapa Bridge at the southern end of the Valley of Tehuacán. The highway here is ascending into the Mixteca Alta.

FIGURE 7.12 San Miguel Tequixtepec, Oaxaca, eastern Mixteca Alta. Scars of ancient terraces are visible to the right of center.

derived from volcanic ash and forming a resistant cap, with very low water conductivity and fertility. In the Mixteca Alta, tepetates vary from whitish to reddish and are equally found in outcroppings or underlying soils. Their presence is a challenge to agriculture. The Mixtecs were able to solve some of its shortcomings through ingenious technologies. Local farmers refer to the deposits as *endeque*, which, unlike caliche, forms a workable rock. That name will suffice, since it is free of previous geological implications.

As the freeway departs the Mixteca Alta toward the south, it winds among foothills into a well-defined oak zone with a smattering of pines (remnants of a once dense pine-oak zone), but the construction of the high-speed highway has left little of the vegetation accessible from the tarmac.

If travelers (wisely, in our view) choose instead to arrive in Oaxaca from Tehuacán using the old, slow Route 135 via Coxcatlán, Teotitlán de Flores Magón, and Cuicatlán in La Cañada, they will find their patience rewarded with a fine dose of forests of cacti and *cuajiotales*, *Bursera*-dominated woodlands, interwoven among fantastic canyon sides that form a folded and faulted landscape. Soon the highway passes by orchards of avocado, Mexican limes (*limón*), mango, and chicozapote along the Río Grande in the canyon country. Side roads offer the chance to explore little hamlets and quaint villages full of archaeological, cultural, and botanical surprises. At the settlement (once a town) of El Chilar, the highway departs from the Río Grande, following its large tributary, the Río de las Vueltas, so named for its many crossings upstream, into the canyon known locally as Cañada Chica. Where water is permanent and raging floods have not completely washed away the riparian forests, ancient giant bald cypresses (*Taxodium mucronatum*), with their prodigious root systems, soften the banks and river edges. Winding into the foothills toward the south, the old highway takes leave of the Cañada Chica at the ancient hamlet of Santiago Dominguillo (formerly Alpizagua) and ascends first into cactus- and legume-rich thornscrub, followed by the short trees of *Bursera*-rich tropical deciduous forest, then into the chaparral complex, vegetation of great diversity, rich in cacti (notably the gigantic *Mitrocereus fulviceps*; see figure 2.7), junipers, oaks, agaves, palms, *Nolinas*, and well-concealed cycads, in a matrix of dense shrubs and short trees. Finally, the highway crests into a pleasant land of oaks and pines, all the while passing through rustic agrarian hamlets. Rzedowski mentions that Mexico is a world center of diversity for oaks (about 160 species) and notes that the oak forests of the Mexican tropics are zones of extraordinary plant diversity, with exceedingly complex floristic, ecologic, and structural traits.[25] The oak-pine zone along Mexico 135 also includes some of the fairest landscapes southern Mexico has to offer (and avoids some of the huge swaths cleared of nearly all native vegetation). Milpas and agave plantations for producing pulque and mezcal alternate with woodlands of oak and pine. From this rich habitat, reaching over 2,000 meters in elevation, the highway departs the drainage of La Cañada as it winds downward into the ancient town of Telixtlahuaca (known locally simply as Télix) and a few kilometers later descends gently into the Valley of Etla. Slightly beyond that fair landscape lies the crowded Valley of Oaxaca.

Prehistoric and Historic Alterations in the Landscape: The Mixteca Alta

Humans have been present in the valleys of Cuicatlán and Tehuacán for more than ten thousand years and in organized semiurban or urban centers for at least three millennia. Their numbers have been generally on the increase during the last 1,500 years, except for a century or so following the Conquest, a time of population collapse due to disease and other vicissitudes. The landscape shows the effects of dense population, especially in the Mixteca Alta. Even the pre-Columbian condition of the valleys would have exhibited the footprint and handiwork of humans. For

25. Rzedowski 1978.

FIGURE 7.13 San Juan Bautista Cuicatlán and La Cañada, viewed from the south. Note the narrowness of the valley, compared with Tehuacán Valley, as seen in figure 6.17. Upstream, the valley narrows into an open canyon, or cañada. Groves to the right of center are primarily mangos and chicozapotes.

centuries they carved and terraced the hillsides to create milpas. They harvested vast tonnages of firewood for cooking, firing pottery for domestic and artistic use, and drying salt in kilns. Perhaps the greatest demand for firewood came from roasting limestone to produce lime mortar for creating stucco and for holding great blocks of stone in place in the hundreds of monuments that frequent the valleys and uplands of the Mixteca Alta.[26] Pre-Columbian workers extracted timber from the forests, cleared forest land to grow food for a large population, imposed dams and check dams on watercourses and drainageways, and carved out canals to control the flow of water for their purposes. The economic demands of organized society and the material enrichment of the elite classes cast deep shadows into the contemporary Mesoamerican monte.

---

26. Oaxacans distinguish among the Mixteca Alta, defined by the cities of Nochixtlán and Tlaxiaco; the Mixteca Baja, or *ñuiñe*, defined by the region surrounding Huajuapan de León; and the Mixteca de la Costa, the region around Putla de Guerrero, Pinotepa Nacional, and south to the Pacific Coast. Several non-Mixtec peoples are found in each division. For an excellent description of the geography of the Mixteca region, see Diguet 1906, 16–19.

FIGURE 7.14 Sabinos (*Taxodium mucronatum*) lining the arroyo in the Río Tomellín drainage, Santiago Huauclilla, Mixteca Alta, Oaxaca. The trees, with rust-colored foliage at this time of year, grow very large and are of great age. Their roots penetrate deeply into the streambed or bedrock. Trenches to the left are recent. Those less visible on the right are probably ancient.

FIGURE 7.15 Oak-pine complex, Nacaltepec, Oaxaca. This is an extraordinarily diverse vegetation. The numerous oak species share space with the palm *Brahea dulcis* and two pine species, plus a host of bromeliads, *Nolinas*, and agaves, among others.

FIGURE 7.16 Forest of a nearly pure stand of oaks near Apoala. The forest is frequently interrupted by cleared plots planted with corn.

How the landscapes would have appeared prior to the invasion of European grazers (horses, mules, donkeys, goats, sheep, pigs, and, yes, chickens) is difficult to ascertain. Although the valleys have remained largely stable geologically since the Middle Tertiary, recent studies reveal a much greener coverage from the Late Pleistocene into the Holocene (2 MYA–present).[27] Lush forests and freshwater lakes and lagoons were common, along with an abundant large mammal fauna, including bison, camelids, glyptodonts, horses, and mammoths, most of which frequented the landscape until roughly twelve thousand years ago.[28] With the appearance of larger groups of humans, the megafauna—including megaherbivores—disappeared.[29] European introductions had an immediate and dramatic effect on the landscape, far more profound than any we can document in prehistory.

Based on vegetation remaining in portions of the Mixteca Alta, we can speculate that the original vegetation (i.e., prior to dense human habitation and livestock grazing) in the vast eroded uplands would have been predominately oak-dominated woodlands intermixed with badlands like those that now cover much of the area. Some of that primordial vegetation will probably never return to its

27. C. E. Smith 1965b, 140–41.

28. Guerrero-Arenas, Jiménez Hidalgo, and Romero 2010.

29. See Martin and Klein 1984, 354–403.

former condition. Other portions, particularly in the upper reaches of the canyon country east of Nochixtlán and Coixtlahuaca, support chaparral that on superficial examination appears depauperate. It comprises a mix largely of grasses, a few cacti, yuccas, *Nolinas,* low shrubs, and small trees, especially junipers, that seldom form a canopy. On closer examination, however, these often-broad expanses present surprising diversity and often grade quickly into well-developed oak forest. In this confusing array of vegetation regimes, one sort of vegetation merges into another in a short distance, now resembling chaparral, now thornscrub, now juniper-rich scrublands, now grassland, now tropical deciduous forest. Portions of it are probably little altered from their pre-Columbian and pre-Mixtec condition. With the arrival of humans, the agroecosystems originated as complex new ecological communities comprising domesticated crops, semidomesticated plants and animals, and ruderals, plants typical of pioneer ecosystems with the ability to colonize disturbed and agricultural land.[30]

Erosion and Prehistoric Control Measures

Northern Oaxaca, the Mixteca Alta in particular, exhibits the most severe erosion of soils in the Republic of Mexico. What proportion of this phenomenon is naturally occurring and how much is anthropogenic is the subject of much controversy. Nearly all analysts agree, however, that humans have contributed greatly to the erosion and, perhaps, to control of erosion as well. The stretch of Highway 135D between Nochixtlán and Coixtlahuaca is especially revealing of the erosion in the Mixteca.

As the incorporation of domesticated crops permitted (or increasingly required) settled pre-Columbian societies, early agriculturalists altered the landscapes, which slowly assumed features still visible, especially on the hillsides. Archaeologists have found that the pressure for greater food production to feed the ever-increasing population led to the widespread construction of hillside terraces, or *lama bordos,* also called *jollas.* These were thick cross-drainage retention dams and buttressed terraces that trapped sediments. At the same time they provided rich agricultural soil as well as checking erosion and preserving moisture. Known as *coo-yuu* in the Mixtec language, the structures permitted vast expansion of agricultural productivity in the Mixteca Alta as well as expansion of the economic and military power of the Mixtecs. David Leigh and colleagues found evidence of *lama bordos* going as far back as 3,400 years.[31] Human intervention through the widespread construction of these lengthy walls and check dams controlled and even reversed natural erosive

30. *Ruderal* is derived from the Latin *rudus* = rubble.

31. Rivas, Rodríguez, and Palerm 2008, 6–16; Leigh, Kowalewski, and Holdridge 2013.

FIGURE 7.17 Modern terraces, Nochixtlán, Oaxaca. The slopes and soils in this region were severely degraded over the centuries. These structures in arroyo bottoms imitate the ancient *lama bordos* of the Mixtecs, serving as both water control and soil-stabilizing structures.

patterns during the peak periods of their use.[32] At least one researcher believes that pre-Columbian farmers in the Mixteca Alta deliberately denuded slopes above their *lama bordos* to accelerate erosion and thus gain the fertile clays that washed down and were trapped by their terraces and check dams.[33] In early descriptions of the Mixteca, Spaniards express the wonder they experienced when first encountering the vast number of *lama bordos* and their remarkable productivity.[34]

Traditional farming in the Mixteca Alta was severely disrupted by the Conquest. Imported lethal infections, epidemics of diseases mostly of European origin, decimated the human population and disrupted the maintenance of the terrace and check-dam system.[35] Without regular maintenance by communal work gangs, the

32. Spores 1969; Kirby 1972, 36; Flannery 1983, 330; Spores and Balkansky 2013, 109; Pérez Rodríguez and Anderson 2013.

33. Flannery 1983, 330.

34. See Burgoa (1674) 1934, 275.

35. A population collapse to less than 10 percent of the original population density occurred in some areas. The specific infective agents are elusive. While most population losses were due to epidemics of diseases of European origin, in some cases the lethal disease may have been a hemorrhagic fever native to the Americas that

FIGURE 7.18 Caliche and tepetate landscape near Coixtlahuaca. Abandoned *lama bordos* are visible, while modern structures are being excavated. Surviving soils are highly friable.

lama bordos fell into disrepair. Spanish encomenderos also exacted heavy tribute payments from Mixtec communities, perpetuating the colonial system the Aztecs had instituted. At the same time Indigenous communities discovered new sources of revenue by producing silk and wheat (mainly for consumption by Spaniards), as well as by acquiring *mercedes*, permission from Spanish authorities to raise herds of goats and sheep.[36] The new enterprises made payment of tribute to the encomendero less burdensome, but they also diverted labor from maintaining the *lama bordos*, which in turn undermined the traditional Mixtec economy. Eventually, especially with the introduction of livestock, erosion tore away the structures faster than the remaining population could maintain them, thus producing or adding to the desolate landscape we see today in much of the Mixteca Alta and perhaps in the Valley of Zapotitlán as well.

The underlying cause of the gradual (or, perhaps, relatively sudden) agricultural decline of the Mixteca Alta lies in its geological setting, the fragile nature of its soils,

wiped out entire Indigenous communities while sparing most Spaniards. See Acuña-Soto, Calderón Romero, and Maguire 2000; and Warriner et al. 2012.

36. Terraciano 2002, 234–36; Burgoa (1674) 1934, 279, describes a flourishing silk industry in the mid-seventeenth century.

and its dependence on *lama bordos* (*jollas*) for productivity.[37] Though much of the land is fertile, many of the soils are low in clay, lacking in the ability to clump and thus unusually susceptible to erosion by wind and water. Any removal of the vegetation cover would expose the soils to rapid transport. With accelerated erosion, other soils expose the formation of caliche, tepetate, and *endeque*, or hardpan.

By the early seventeenth century, Spaniards had settled in Oaxaca in considerable numbers, both in the valley bottoms, where soils were richest and water easily available, and in the Mixteca Alta, with its Castile-like climate and fertile soil. The aridity of the Valley of Tehuacán was far less appealing to the European transplants than the verdant central valleys and the cool, moist uplands of Oaxaca. Early accounts immediately following the Conquest describe a prosperous, agriculturally productive region in the valleys of Oaxaca, but notably in the Mixteca Alta as well, where every hill and mountain was graced with tier after tier of terraces, hundreds of kilometers of them, planted with corn and other crops.[38]

Spaniards also discovered the numerous valleys of the forested Mixtec uplands to be thickly populated enough to warrant the investment of a host of clergy eager to lead the Indigenous inhabitants out of what they perceived as their dark ignorance of spiritual reality. These natives were sufficiently numerous to provide labor to construct a wide array of convents and churches, and by the mid-sixteenth century, widespread building was under way. The newly rising structures ranged from the massive Dominican convents at Teposcolula, Yanhuitlán, and Coixtlahuaca—the latter visible today from the high-speed expressway—to lesser but still impressive parish churches, a minimum of one in every village and town. The Spanish Conquest brought with it the imposition of the *repartimiento*, a system imposed by the Crown whereby communities were required to provide gangs of native construction workers. The labor requirement was added to the already traumatic imposition of European systems of agriculture on the vanquished natives.

European agricultural practices quickly led to disruptive alteration of the landscape and of the regional economy. The key elements of this newly imposed agrarian system were the introduction of European livestock and the imposition of wheat culture. Spaniards also introduced silkworms and planted mulberry trees to feed them, but this introduction appears to have posed no great effect on agricultural production.

Of the two major introductions, the arrival of livestock instigated the far greater change to the landscape and the economy. Prior to the Conquest, only dogs, turkeys, Muscovy ducks, stingless bees, and perhaps rodents were domesticated. By the mid-sixteenth century cattle, sheep, and goat farming were entrenched in Mesoamerica

37. Krasilnikov et al. 2013, 167–68.

38. Burgoa (1674) 1934, 275.

FIGURE 7.19 Combination of modern and ancient terraces, Yucuita, Mixteca Alta.

and became the leading source of income for the Mixteca Alta. So successful was this new industry (primarily for Spaniards) that in only a few decades, the uplands of southern Mexico had become irreversibly overgrazed, the soils heavily compacted by trampling, and the hillsides exhibiting the early effects of erosion. In 1560, the viceroy of New Spain, don Luis de Velasco, learned of the impoverishment of the natives of Oaxaca. Velasco had received a deluge of reports from throughout New Spain of the damage to Indigenous crops and land wrought by cattle. In response, he ordered cattle removed from native lands and relegated them to lands on the periphery. His orders did not extend to sheep, goats, mules, burros, and pigs, however, and these continued to denude and trample grasslands and uplands, as well as devouring crops.[39] Besides ordering the removal of cattle, he provided the native peoples with *mercedes* and *estancias*—grazing rights for goats and sheep—to alleviate the problems engendered by open range grazing by the Spaniards' livestock herds. For the Mixteca, between 1560 and 1620, the Crown granted many such mercedes for the communal use of towns, for the personal use of Mixtec caciques, and in lesser amounts, to Spaniards. The latter cannot have been pleased by the viceroy's edicts favoring the rights of Indigenous Mixtecs.[40]

39. Melville 1994, 136–38; Mendoza García 2002, 758.

40. Mendoza García 2002. For many Spaniards who had been landless or peasants in Spain, owning a cattle ranch was a prestigious mark of social elevation, especially if they no longer had to work. At the time of

FIGURE 7.20 Courtyard of the mid-sixteenth century Coixtlahuaca convent. The huge quantity of mortar and plaster necessary for construction of the convent required stupendous amounts of labor and roasting of limestone, taxing local wood supplies, already heavily harvested for cooking and building.

Whether cattle were banned completely from the populated portions of the Mixteca Alta is unclear, but grazing by *ganado menor* (primarily sheep and goats) continued unabated, with disastrous long-term consequences. The Mixtecs discovered that livestock could become an economic resource, which, along with silk production, had not existed in pre-Columbian North America. In a short period Indigenous people became pastoralists in addition to farmers.

In addition to sponsoring the livestock invasion of the Mixteca Alta shortly after the Conquest, Spaniards introduced wheat cultivation into the Mexican highlands, eager to duplicate their European diet and provide the clergy with liturgical bread. The climate and soils proved ideal for wheat cultivation. It was first planted in the Atlixco Valley, not far from Tehuacán, and from there spread rapidly to the north and west of Mexico City.[41] Before long, it arrived in the Mixteca Alta as well.

Viceroy Velasco's death, Crown investigators determined that he had died in poverty due to his untiring efforts to bring justice to the native peoples of New Spain, often opposing the claims of Spaniards to Indigenous land.

41. Bakewell and Holler 2010, 241.

FIGURE 7.21 Commons grazing near Coixtlahuaca. For pastoralists, sheep such as these represent their bank account. The agaves are probably *A. salmiana*, the pulque agave. In the region, such plants are a source of income and are maintained and monitored.

FIGURE 7.22 Wheatfields near Tepelmeme de Morelos, Mixteca Alta. In this region whole wheat tortillas are more commonly produced than those of corn.

The agronomy of wheat is dramatically different from that of corn. In this region, with the advent of wheat, which supplanted corn over a wide area, the diversity of multicropping diminished. Wheat is intolerant of competitors; hence it is planted alone. Corn is usually planted in the company of the other two members of the famous Mesoamerican Three Sisters, beans and squash. Wheat is more sensitive to drought than corn and requires routine fertilizing, more so than corn, for milpas are naturally enriched with nitrogen fixed by the roots of companion beans. As wheat culture expanded, soil fertility likely decreased by that lack of nutrient inputs and field rest and fallow. The soil protection provided by the stubble from milpa crops of corn, beans, and squash was drastically decreased as well. Ironically, in the eastern Mixteca Alta, the best fields, those in the valley bottoms, still tend to be dedicated to wheat, while corn is often relegated to the eroded upland slopes, *jollas*, and some rain-fed milpas.

The widespread culture of wheat in the Coixtlahuaca region, a curious inheritance from the Dominicans, is nearly unheard of elsewhere in Mesoamerica south of Mexico City. In Coixtlahuaca and Tepelmeme de Morelos, a former Chocho town some 20 kilometers north, whole wheat tortillas have replaced corn tortillas as the preferred dietary starch. These tortillas, however, are not the dietary nightmare of Mexico's north, which feature white flour, lard or shortening, and salt. The Mixteca Alta variety contains only whole wheat and salt, no added fat whatsoever. Farmers in the region long ago became adept in raising wheat and selected varieties adapted to Mixteca's climate, which flourish, and Indigenous folk now plant wheat in plots, often alongside their milpas of corn. Tepelmeme is home to three powerful grain mills, which operate for several hours each morning. The housewives take their wheat each day to be ground into flour at the mills.

To fashion tortillas, the women wash the flour to concentrate gluten, add water and salt, and pat out the tortillas to a diameter of around 40 centimeters (16 in.), then toast the tortilla on a large comal (griddle). The finished product equals in size the finest of flour tortillas from the north, but without the refining of the flour and addition of hydrogenated fats commonly found there. The remainder of the Chocho diet is like that of Oaxacans everywhere, rich in chiles, squash, and beans.

Along with livestock, wheat, and European diseases came the gradual abandonment of the *lama bordos* (*jollas*). These sophisticated structures fell into a state of disrepair, slowly eliminating a major source of agricultural productivity and inhibitor of soil erosion. The new economics based on livestock provided revenue, at least in a sizable part, to newly established Catholic missions for the construction of the great convents and churches. The cost of diverting much of the small remaining population into laborers for church construction was the neglect of the maintenance of the *coo-yuu*, the vast and complex network of terraces and *lama bordos*. The demise

FIGURE 7.23 Chocho woman with whole wheat tortilla, Tepelmeme.

of those structures was followed swiftly and inevitably by the rampant erosion that had already been compounded by the introduction of livestock. The *lama-bordo* system required cooperative efforts, and any loss of labor force or diversion into non-farm activities immediately jeopardized the integrity of the system. To some extent (the degree is controversial) the frightening erosion now apparent in the Mixteca Alta began with the abandonment of the *lama bordos*. Since late in the twentieth century, the Mexican government has attempted to address the catastrophic rates of erosion by major campaigns of reforestation and erosion control.

In the hotter lowlands of the valleys, human attempts to cope with erosion and chronic scarcity of water have assumed different forms. Shortage and variability of rainfall were perennial conditions that entailed a constant need to capture rainwater, construct water conservation and management projects, and incorporate native plants into a subsistence way of life. The pre-Columbian Valley of Tehuacán was lined with hundreds of kilometers of canals arranged in dendritic formation.[42]

42. Woodbury and Neely 1972, fig. 151.

Presa Purrón is a powerful example of Late Formative pre-Columbian investment in water control. Two large canals in the valley demonstrate the creativity of native engineers and hydraulic experts in the Valley of Tehuacán: one built during Classic times, a brick-lined structure 5 kilometers long in the far south of the valley, the other from the Postclassic, 25 kilometers long, conducting water from near Tehuacán to south of Miahuatlán (see figure 4.12). Both of these expanded the agricultural potential of the valleys by making irrigation water available to fertile but arid fields.[43] In the Cuicatlán Valley, as Eva Hunt has described, a complex social and physical network of water claims and delivery systems mark the prehistory of La Cañada.

Into the Valley of Zapotitlán

Nowhere is the challenge of arid lands better exemplified than in the Valley of Zapotitlán. In our earliest travels along Mexico Route 125 to Oaxaca from Tehuacán via Huajuapan de León, instead of the Highway 135D or old Route 135, we were struck by the aridity of the landscape almost immediately.

Heading southwest from Tehuacán, once one leaves the valley behind at Santa María Coápam and heads under the freeway a few kilometers southwest, the landscape becomes starkly different. The road climbs into arid hills of white, stony lithosols and poorly developed calcareous soils that are too steep and arid for agriculture. Strange plants and huge cacti stamp this landscape as desertlike.

At Texcala, a rock-carving center a few kilometers of winding highway into the hills, another feature can nearly cause one to drive off the winding mountain road: tier after tier of terraces and tunnels, ancient in origin, some of them used now as homes or garages. These ancient structures were painstakingly constructed and maintained over the centuries for producing salt and housing the resident salt producers. Now the salt seeps within the town have gone dry or are greatly diminished and the inhabitants carve figurines from a rock they call *ónix*. It is actually travertine, a rock formed from water seeping through limestone. A couple of kilometers farther on, in a canyon bottom, lie active salinas, a rainbow network of the flooded salt pans for which Zapotitlán has been renowned for centuries and which still yield marketable salt. Within a 15 kilometer radius lie many hundreds of carefully leveled pans, each worked by a salt producer who waits patiently for the sun to evaporate water from the shallow pans, leaving behind the precious salt. Only in an arid environment can salt be produced by evaporation.

43. Woodbury and Neely 1972, 130–31.

FIGURE 7.24 Desert landscape near Texcala, Puebla, Valley of Tehuacán. The large cacti are *Mitrocereus fulviceps*, the yuccas, *Yucca periculosa*. Just above center to the left is a large *Beaucarnea gracilis*.

The winding, narrow roadway splits cactus forests of eerie density near Zapotitlán Salinas. A careful examination of the hillsides around reveals that this arid valley contains the remnants of ancient terraces by the hundreds, now virtually all abandoned, but once, long ago, justifying and rewarding the labor required for their construction and maintenance.[44] In the Zapotitlán Valley, fertile soils barely exist. The basic ground material is an alkaline, shallow substrate that will produce crops only with intensive management. Building the terraces that would produce corn, beans, and squash required a heavy investment in labor over many decades and knowledge of dealing with highly alkaline substrates.[45] These terraces are proof that through human intervention and manipulation, the arid landscape once produced harvests of corn, beans, and squash.

44. See Woodbury and Neely 1972, 120.

45. López-Galindo et al. 2003, 19.

FIGURE 7.25 Texcala, Puebla. The town's industry is carving figures and artifacts from travertine, labeled *ónix* by the artisans. The buildings are constructed on terraces formerly housing salt pans or remnants of travertine mining. In the far upper left corner, numerous *Mitrocereus fulviceps* blanket the hills.

FIGURE 7.26 Salt works, Zapotitlán Salinas. This is a modern operation using pre-Columbian structures. (See also figure 6.4.)

FIGURE 7.27 Plantation of *maguey pulquero* (*Agave salmiana*), the principal source of pulque. This field is at about 2,000 meters elevation, ideal for production, but situated on a ridge roughly 400 meters above Zapotitlán Salinas and accessible only on foot.

FIGURE 7.28 Tlachiquero extracting aguamiel, the sweet liquid exuded from the center of the pulque agave, using an *alacate*, also called *acocote*. Once aguamiel is collected, it rapidly ferments into pulque.

Zapotitlán Salinas was probably founded by Franciscan friars, who forced the people of the citadel atop Cerro Cuthá, 2 kilometers east, to relocate to a site more conducive to conversion to Christianity and supervision by the clerics. In much of the twentieth century, Zapotitlán's economy was based on agrarian resources (raising goats) and the revenues from the salt pans. Today much of its population of around two thousand depends on commerce and tourism, remittances from relatives living in the United States, and other more obscure sources of revenue. Pulque, the fermented juice gathered from the agave called *maguey pulquero* (*Agave salmiana*) is available daily for much of the year, transported burro-back from plantations high in the Sierra Zapotitlán. The popularity of pulque represents a sure indication that the town retains its ancient roots.

Vegetation of the Valleys

Within their surface area of about 10,000 square kilometers (3,861 sq. mi.), the valleys of Tehuacán and Cuicatlán and their environs comprise an ever-changing assemblage of plants, a collection so different from anywhere else that biogeographers consider it a separate floristic province. Both of us have studied the flora and vegetation of the Río Mayo region of northwest Mexico, an area of renowned floral diversity, with a mix of thornscrub, tropical deciduous forest, oak woodland, and mixed conifer forest, among others. As of the year 2000, in the roughly 40,000 square kilometers, researchers have cataloged 2,825 plant species, an extremely high number. In comparison, the valleys, roughly one-fourth the size of the Río Mayo region, have nearly the same number of plant species in an even greater variety of habitats. Jerzy Rzedowski includes the Tehuacán and Cuicatlán Valleys as the southernmost outpost of the Mexican Xerophytic Region, which encompasses most of central and northern Mexico, the U.S. Southwest, and Texas.[46] These two arid valleys and their side drainages occupy less than 3 percent of Mexico's landmass but are home to nearly three thousand species of identified vascular plants, about 10 percent of the Mexican flora, making the region one of the biodiversity hotspots of the Americas. In the words of Rzedowski, Mexico's topography is like a crumpled piece of paper where humid high mountains rise from low arid drainages and gorges. The Tehuacán and Cuicatlán Valleys, tropical in geographic location, exhibit an exceptional environmental diversity created by the elevational and rainfall gradients, which are key elements promoting species diversity. The numbers include about 180 plant families and almost 900 genera. Angiosperms—flowering plants—are

46. Rzedowski 1978.

the dominant group, with more than 2,600 species. Plants include almost 60 moss species, surprising in a climate of such aridity, more than 150 fern species, and 20 gymnosperms—including 4 cycads. More than 350 species (14 percent of the flora) are endemic—found only in the valleys.[47]

In addition to plant diversity, the valleys also demonstrate a striking ethnobotanical heritage. The long human occupancy of the valleys is evident not only through the noticeable land use change over time, and the multitude of milpas and terraces (mostly abandoned), but also by the presence of 815 plants used throughout the valleys. These include 685 wild plants, 167 cultivated, 117 weed and ruderal (growing in waste places), and 59 managed in situ plants.[48] This vast compendium of useful plants is a clear indicator of a high degree of plant diversity and of long-term human occupation.

Most visitors will easily identify the diversity of the ever-changing vegetation where unique, and sometimes grotesque, forms dominate the landscape. Ecologists recognize as many as thirty-six different vegetation units within the boundaries of the valleys.[49] These range from mesquite woodlands on riverbanks to complex associations of giant columnar cacti with trees and shrubs in the lowlands, and from diverse oak forests to shrubland and chaparral complex plant communities—sometimes called *mexical*—on higher ground, and from tropical deciduous forest to depauperate caliche flats. For purposes of simplification, these many types of vegetation can be reduced to a few basic vegetation regimes:

1. *Thornscrub*, which includes cactus-rich forests, the savanna-type lowlands of the western sides of the valley, and several related plant associations, especially those of succulent and sarcocaulescent trees. Examples of cactus-rich thornscrub are groves locally called *tetechera, cardonal, jiotillal,* and *chichipera*.[50]
2. *Tropical deciduous forests* (TDFs), characterized by winter or dry season deciduous trees that produce new leaves in response to the strongly seasonal summer rainfall and drop their leaves in response to fall drought. TDFs comprise various types, some including a strong presence of columnar cacti, others with different associations of trees, especially *Burseras* and legumes.
3. *Palm forests/scrub habitats*, where palms tall and short overwhelm other plants. This habitat varies from that of palm groves with its relatively open understory to lush shrubbery and dense, impassable low-stature thickets.

47. Dávila et al. 2002; Dávila et al. 1993–2012.

48. Casas et al. 2001.

49. Valiente-Banuet et al. 2009, 32–35. This is a useful introduction to vegetation of the valleys.

50. *Tetechera*: forests of *tetecho, Neobuxbaumia tetetzo*; *cardonal*: forests of *cardones, Pachycereus weberi*; *jiotillal*: forests of *jiotilla, Escontria chiotilla*; *chichipera*: forests of *Polaskia chichipe*. In most cases the word *forests* is a stretch. *Groves* or *stands* would be more appropriate.

4. *Chaparral complex,* a vegetation commonly found in elevations intermediate between TDF/low thornscrub and oak woodland. This includes the sui generis plant community that swathes the steep hillsides of the Cañada Chica above about 1,000 meters elevation. It also includes upland thornscrub, *mexical,* and shrubland plant communities, as found in the higher elevations of some slopes in the Valley of Zapotitlán and eastern slopes of the Valley of Tehuacán.
5. *Oak forests,* ranging from very dry and sparse (more a type of arid woodland) to a closed canopy of tall trees, a high diversity of tree species and epiphytes, and a mesic environment.
6. *Pine, pine-oak, and alder forests,* prevalent only in the high-elevation sites.
7. *Highly specialized vegetation types,* such as riverine systems with gallery forests of mesquite and willows in the arid lowlands; gallery forests with many tropical species in the tropical, more mesic lowlands; and gallery forests of towering Montezuma cypresses in the smaller, permanent drainages on mid-elevation sites.

Thornscrub and Its Allies

Cactus-rich thornscrub is the charismatic vegetation that has made the valleys famous. It is by far the most prominent regime of the valley floors and bajadas, especially of the Valley of Tehuacán and, even more so, its tributary Valley of Zapotitlán. This is a highly varied vegetation type. It presents a seemingly unending novelty of plant associations in response to subtle changes in the ecological setting. Most of the valleys' unusual plants and plant associations occur within this grouping of plant communities and habitats. These appear as vast stretches that support small to medium-sized acacias, mesquites, mimosas, *Parkinsonia praecox, Cyrtocarpa procera,* and, conspicuous, especially in the southern portion, *Fouquieria formosa.* In much of the lower valleys, columnar cacti tower above the often-sparse shrubby vegetation.

In other locations at elevations above roughly 1,400 meters in the valleys of Tehuacán and Zapotitlán, cacti are less apparent. Instead, sporadic clusters of izote (*Yucca periculosa*) or *sotolín* (*Beaucarnea gracilis*) appear on the landscape, rising with their distinct shapes above what appears to be undifferentiated low shrubbery or nearly bare soil. This vegetation is seldom far from flocks of goats or solitary burros, both of whose foraging habits have shaped the vegetative landscape over the last five hundred years.

The Cuicatlán Valley, though parts of it are as dry or drier than the Valley of Tehuacán, gives the appearance of thicker or denser vegetation and a higher percentage of ground cover. Much of this is due to soils, for although rainfall in the lower La Cañada is even less than that of much of the valley of Tehuacán, the stark caliche

FIGURE 7.29 Thornscrub in the dry season near Santiago Dominguillo, La Cañada. Yellow flowers are of *Parkinsonia praecox*. Columnar cacti include *Pachycereus weberi* (bluish), *Escontria chiotilla*, *Myrtillocactus geometrizans*, and *Stenocereus pruinosus*.

FIGURE 7.30 Columnar cactus forest—primarily *Neobuxbaumia mezcalaensis*, San Juan Raya. David Yetman stands for comparison.

FIGURE 7.31 *Izotal* (thicket of izotes, *Yucca periculosa*), Zapotitlán Salinas.

of the valley to the north, especially in the Valley of Zapotitlán, is far less prevalent in La Cañada. Somewhat higher humidity may also be present in La Cañada, associated with the breach through the Oaxaca Fault near Quiotepec, which allows moist air masses to penetrate from the Gulf coast, rising along the Río Santo Domingo. La Cañada is also far narrower, the hillsides steeper, and the Río Grande flowing with far greater volume, its waters less saline than those of the Río Salado. And rainfall appears to increase rapidly as the steep slopes rise from the valley bottom, dramatically more so than in the Valley of Tehuacán. These factors help account for the greater plant density and (often) size along with the richness of the vegetation in La Cañada. In addition, La Cañada is lower and thus more tropical than any other part of the valleys. The effects of the warmer climate are observable in the rich associations of *Bursera* species and columnar cacti on the hillsides of the lower Cañada Chica and La Cañada, extending into the canyon sides of the Río Santo Domingo. These associations reach well north into the lower valley of Tehuacán as well, notably on the plains in the vicinity of Cerro Petlanco.

Vegetation of the western side of La Cañada is rather drier than on the eastern side. Dry lands extend almost all the way from the valley bottom into much of the eastern Mixteca Alta. There, in the highlands above La Cañada and the southern

FIGURE 7.32 Diverse thornscrub, west side of La Cañada. Cacti are *Cephalocereus columna-trajani* and *Neobuxbaumia tetetzo*.

valley of Tehuacán, we find broad areas of severe and widespread erosion and extensive tracts of land devoid of forest vegetation but intermixed with a chaparral complex of surprising diversity.

The Zapotitlán Valley exhibits vast expanses of thornscrub, a terrain of harsh aridity, an eroded landscape, in part at least the result of being heavily grazed for centuries. Part of its attraction are the broad swaths of utter dominance by cacti, *Nolinas*, yuccas, and thorny shrubs. Even at the summits of the smaller mountains—the Sierra Zapotitlán, as it is locally called—and the northern foothills of the Sierra Mixteca, trees are mostly of small stature, which makes the profusion of columnar cacti and yuccas and their relatives even more visually spectacular. Throughout the reaches of the valley, abandoned milpas, outlined by rock terracing, demonstrate the decreasing agricultural productivity of the fields over the centuries.

Despite the parched landscape, the valley's extraordinary flora shines through. With a little patience and willingness to venture off onto dirt roads north and south of the pavement, the traveler can locate all sixteen species of the columnar cacti that frequent the Valley of Zapotitlán. An additional two species, *Pachycereus grandis* and *P. weberi*—low-elevation columnar cacti species—are found in the Valleys of Tehuacán or Cuicatlán but do not appear in the Valley of Zapotitlán.

We have given a cursory description of the highway journey from Tehuacán into the Valley of Zapotitlán. We now present it in more detail, since it is this valley that has caught the attention of botanists, especially cactologists, for over a century.

Starting from Santa María Coápam, a Nahua village at the western edge of the valley, now a suburb of Tehuacán, Mexico Route 125 quickly winds into arid hillsides toward Zapotitlán Salinas. One enters first the zone of towering *órganos* (*Mitrocereus fulviceps*), clumps of tall izotes (*Yucca periculosa*), and lumpy, massive biznagas (*Echinocactus platyacanthus*; see figure 7.24).

Soon the roadway descends through Texcala and abruptly enters the zone of *viejitos*, the endemic *Cephalocereus columna-trajani*, thousands of them, resembling curved-shouldered soldiers ascending the steep hillsides. All of them show orderly tilted stems and northward-positioned *pseudocephalium*—the dense woolly and bristly portion of the stem that protects the reproductive structures and often extends down the thick stems from the apices. *Viejitos* appear to sprout branches only in response to injury. Stem tilting and pseudocephalium orientation have functional advantages in terms of how the plants intercept direct solar radiation. The angle of tilting and the location of the pseudocephalium vary latitudinally in columnar cacti, imposing mechanical restrictions to their height.[51] This species may have evolved in the valleys with a genetic predisposition to the angle in which they intercept sunlight, sharply limiting their ability to grow north or south of their present range.

The sinuous highway descends through hills of seemingly endless numbers of the *viejitos*, past the salt works. Then, just as suddenly, the *viejitos* merge into the forests of *tetechos*, *Neobuxbaumia tetetzo*, generally skinnier than the *viejitos*. Their numbers are vast and their density such that they form true forests. The Mexican government recognized the spectacular nature of the forest and established the Helia Bravo Hollis Botanical Garden, complete with guides of Popolocan extraction. The garden covers a gentle hillside in the shadow of the adjacent Cerro Cuthá, where ancient Popolocas built a monument-covered citadel now overgrown with thornscrub and tropical deciduous forest.

Continuing past Zapotitlán, the valley bottom—much of it deeply incised by channel cutting—rises steadily in elevation. On the flats near the roadsides are groves of the sometimes scruffy-appearing columnar cacti, *babosos* (*Pachycereus hollianus*), often used as living fences in corrals and along roads. Ten kilometers or so north of the highway sits the hamlet of San Juan Raya, plunked down, it seems, amid some of the densest forests of columnar cacti in the world and seated on deposits of fossils of extraordinary abundance. This part of the upper valley of Zapotitlán is home to impressive numbers of the columnar cacti *Neobuxbaumia mezcalaensis*,

51. Valverde et al. 2010.

FIGURE 7.33 *Viejitos* (*Cephalocereus columna-trajani*) showing pseudocephalium on the north side of the plants.

FIGURE 7.34 *Cephalocereus columna-trajani*, Zapotitlán Salinas. This cactus is endemic to the valleys. It is named after Trajan's Column, a monument in Rome.

FIGURE 7.35 *Tetechera*—forest of *Neobuxbaumia tetetzo*, Zapotitlán Salinas. Note the strongly calcareous soils. This is one of the world's densest cactus forests. In the foreground are the *sotolín* (*Beaucarnea gracilis*) and izote (*Yucca periculosa*).

perhaps the tallest of all cacti, and *Neobuxbaumia macrocephala*, locally referred to as the *órgano de cabeza roja*.[52] The latter is found only in this area and in a thriving population near San Antonio Cañadas on the eastern side of the Valley of Tehuacán. Combined with eerily weird groves of *sotolín* (*Beaucarnea gracilis*), izote (*Yucca periculosa*), giant biznaga (*Echinocactus platyacanthus*), mounds of the diminutive but many-headed *Ferocactus robustus*, scattered *Burseras*, *Fouquieria formosa*, and several species of agave, this landscape is a botanist's dream world, an unforgettable landscape stretching far to the north into the foothills of the Cerro Gordo, part of the Sierra Zapotitlán (see figure 2.6).

The environs of Zapotitlán Salinas are the best place to view the *sotolín*, the grotesquely swollen-stemmed *Beaucarnea gracilis*, often called the *pata de elefante*, and the colossal biznaga, the barrel cactus *Echinocactus platyacanthus*, jokingly called *silla de suegra* (mother-in-law's seat). These two regional endemics are usually

52. Yetman has photographed an individual roughly 20 meters tall growing in a canyon floor just north of Huajuapan de León, Oaxaca, in the Mixteca Baja.

FIGURE 7.36 *Neobuxbaumia macrocephala* growing near Los Reyes Metzontla, Puebla. The species is endemic to the valleys of Tehuacán and Zapotitlán.

FIGURE 7.37 Cactus forest primarily of *Neobuxbaumia mezcalaensis*, San Juan Raya, Puebla.

FIGURE 7.38 Cristate *tetecho* growth in sixteen years, Los Reyes Metzontla.

FIGURE 7.39 *Sotolines* (*Beaucarnea gracilis*), San Juan Raya. Many more of the trees are visible in the background.

FIGURE 7.40 *Ferocactus robustus*, Zapotitlán Salinas. These peculiar plants appear to reach their maximum growth in the Valley of Zapotitlán.

FIGURE 7.41 *Silla de suegra* (*Echinocactus platyacanthus*), Santa Ana Teloxtoc, Valley of Zapotitlán. They grow slowly, and plants of this size are several hundred years old.

accompanied by tall *Yucca periculosa*, the izote, often growing in dense groves. Izotes appear in stands of eerie appearance in a few locations, reaching nearly 10 meters in height, presenting an otherworldly sensation in the botanically inclined viewer.

A few kilometers southwest, the valley narrows, and the highway begins to wind upward toward the village of Acatepec, entering an enchanting landscape of columnar cacti (fine stands of *Neobuxbaumia macrocephala*); barrel cacti, especially the endemic many-headed, mound-forming *Ferocactus robustus*; several agave species and their relatives; and yuccas in a matrix of leguminous and euphorbiaceous shrubs and small trees.

Southeast of Acatepec on a graded dirt road is the cactus-rich Popolocan community of Los Reyes Metzontla, where native potters carry on an ancient ceramic tradition, and residents plant cuttings of columnar cacti in their yards as a source of fruits. The environs of this town are the best place to view the large candelabriform cacti *Polaskia chende*, *Polaskia chichipe*, and *Myrtillocactus schenckii*, all growing in the same habitat. The graded road continues climbing through Metzontla, winding through hillsides laced with *Beaucarnea purpusii*, growing at a higher elevation than *B. gracilis*, with a trunk not quite so swollen. The winding route passes through the village of Xochiltepec, crosses a divide, and descends to the small town of San Luis Atolotitlán. On the hillsides are fine specimens of *Isolatocereus dumortieri*. Above

FIGURE 7.42 *Myrtillocactus schenckii* (*right*) and *Polaskia chichipe* (*left*).

Atolotitlán to the northeast, on arid hillsides grow hundreds of chichipes, forming a true *chichipera*, quite probably the only one. The road forks in Atolotitlán. The eastern fork leads downstream to Santiago Coatepec. With some diligence and equal luck (and cooperation from protective villagers), the explorer may come upon scattered cycads (*Dioon caputoi*) that have survived poaching. The other fork departs Atolotitlán to the west, leading to Caltepec across the fault of the same name, lying in a different drainage. Along this road are many fine examples of *P. chende* immediately adjacent to the roadway.

Throughout the thornscrub, groves of cacti are often filled with a dense understory of *Mimosa luisana*, a catclaw with viciously recurved thorns that repel livestock—even the toughest goats—and slow the progress of the human intruder. This leguminous shrub appears to be responsible for the recruitment of several species of columnar cacti by providing shelter to the young developing seedlings. The thorns at ground level repel the would-be herbivores, while the shade protects the young plants from desiccation and direct sunlight. At the same time, roots from well-established plants may extract water from moist soils deep underground and share it with the developing seedlings through a process known by ecophysiologists as hydraulic lifting. Nurse plants, as these sorts of nurturing overstory are called, may pay dearly for the goodness and protection they provide, for when the cacti

FIGURE 7.43 *Isolatocereus dumortieri* near Atolotitlán, Valley of Zapotitlán.

reach adulthood, they may outcompete their benefactors, intercepting rain at the surface before it can penetrate to the root zone of the nurse plants. Under the towering cacti—the protégés—it is common to find the dead or dying remains of the nurse plant.[53]

Continuing through Acatepec to the southwest, the valley rises until it reaches the Continental Divide near the state line with Oaxaca. At that point the highway leaves the valley complex, and all watercourses drain into the Pacific. The Río Zapotitlán drains southeast from the western flanks of the basin, entering the Valley of Tehuacán at San Juan Atzingo and, farther downstream, San Gabriel Chilac, then joining the Río Salado not far south of Coxcatlán.

Along with the columnar cacti, some plant communities in thornscrub often feature an abundance of *sarcocaulescent* species—trees and shrubs with swollen stems and branches. Within these associations, different species of the genus *Bursera* are dominant but coexist with columnar cacti, tree morning glories (*Ipomoea* spp.), and the ocotillo relative *Fouquieria formosa*. In communities like these, thornscrub grades into tropical deciduous forest and back again, making categorization murky.

53. Valiente-Banuet and Ezcurra 1991; Valiente-Banuet et al. 1991; Valiente-Banuet, Vite, and Zavala-Hurtado 1991.

FIGURE 7.44 Sierra Zongolica from Zinacatepec, Valley of Tehuacán. The horizontal band of cacti at the base of the foothills is composed of *Pachycereus weberi*. The dense forests above are of *Neobuxbaumia tetetzo*.

Most columnar cacti are residents of thornscrub, but they sort out by elevation, often venturing upward into more mesic plant communities. This is most readily visible in La Cañada, where *Mitrocereus fulviceps* grows in chaparral above *Cephalocereus columna-trajani*, which grows above *Neobuxbaumia tetetzo*, which grows above *Pachycereus weberi*. Starting at the Mixtec village of San Pedro Nodón at 1,700 meters, high above the floor of La Cañada, the altitudinal sorting unfolds delightfully as the roadway descends eastward more than 1,000 meters down into the valley bottom, the succession of columnar cacti evident as one species grades into another. To the north, the Zapotitlán Valley is too high for *Pachycereus weberi*, which prefers the lower parts of the Valley of Tehuacán and La Cañada. *Polaskia chende*, *Polaskia chichipe*, and *Myrtillocactus schenckii* grow at the highest elevations, well over 1,800 meters. The latter three are found together mostly in the vicinity of Los Reyes Metzontla.[54]

On the other hand, the distribution of *Neobuxbaumia tetetzo* in Cañada Chica is puzzling. Dense groves cover hillsides in the valley bottom as far upstream as

54. *Myrtillocactus schenckii* is widespread in Oaxaca east of Oaxaca City, well into the canyon country toward Tehuantepec.

FIGURE 7.45 *Coryphantha calipensis*, Cerro Petlanco.

Atatlahuca, where the *tetechos* abruptly cease. They appear again, however, east of the city of Oaxaca at higher elevations than their limit in Cañada Chica. In contrast with the Valley of Tehuacán, *P. weberi* continue to appear *above* the distribution of *N. tetetzo*. This is their eastern limit. They are absent east of Oaxaca City. *N. tetetzo* seems to prefer more open, rolling country than the ever-narrowing Cañada Chica can provide upstream, while *P. weberi* benefits from the special mesoclimate of Río de las Vueltas, allowing it to climb higher than in the Valley of Tehuacán.

Other columnar cacti are more catholic in their preferences and distribute in a more liberal fashion. For example, *Pilosocereus chrysacanthus* seems equally happy in the oak woodlands, the chaparral complex, or the upper dry lands of the valleys. Scattered among the varied thornscrub habitats are also more than a dozen species of agaves and myriad understory small cacti, including at least twenty-two species of *Mammillaria* (as of 1997), four species of *Coryphantha*, and four of *Ferocactus* (barrel cacti).[55]

Thornscrub associations form a mosaic of plant communities that grade into other plant vegetation types, sometimes into palm forests, at other times into oak forests, or, when water availability is greater, into tropical deciduous forest.

Tropical Deciduous Forest

Mexico has an abundance of tropical deciduous forests, often referred to also as tropical dry forests, or seasonally dry tropical forests. Mexican researchers use the terminology *bosque tropical caducifolio* or *selva baja caducifolia* to refer to these woodlands, which are characterized by often thick growth of short trees (typically 10–15 meters in height) that exhibit sharply different appearance in rainy and dry seasons. Extensive belts of TDF formerly occurred from central Sonora southward along the base of the Sierra Madre Occidental, the Balsas River depression, and coastal western Mexico to the Pacific coast of Costa Rica. They also occur in pockets along the coast of the Gulf of Mexico and in the Yucatán Peninsula.[56] Tropical dry forests also once rimmed major portions of the valleys, especially on north-facing

55. Arias Montes, López, and Guzmán Cruz 1997.

56. Rzedowski and Calderón de Rzedowski 2013, 2, 5.

slopes above the drier parts of the valley floors. Scattered remnants can still be seen throughout the valleys on steep slopes. Specifically, fine tracts are found south of Santiago Dominguillo in La Cañada, in Cañada Chica upstream as far as Jayacatlán, and in Tomellín Canyon as far upstream as El Parián.[57] Excellent stands also remain on western-facing slopes of the Sierra Juárez (also known as the Sierra Cuicatlán) east of Cuicatlán, the Sierra Mazateca east of Teotitlán, and the western slopes of the Sierra Zongolica, places with thin soils, convoluted topography, and little potential value as livestock pastures. An especially rich patch of forest survives within the fenced confines of La Mesa (Tehuacán El Viejo) archaeological site just southeast of Tehuacán (see figure 6.22). On the eastern slopes of the Sierra Mixteca, along the roadway descending from San Pedro Nodón, tropical deciduous forest is restricted to protected canyons and ravines, whereas the more exposed slopes harbor thornscrub or chaparral (see figure 7.32). A remnant of recovering tropical deciduous forest is present on the slopes of Cerro Cuthá, the archaeological site just east of Zapotitlán Salinas.

These habitats are characterized by dense vegetation composed of a variety of trees, mostly shorter than 10 meters, and an abundance of spiny understory, cacti, and agaves. The forest is less widespread than thornscrub and often remote from highways. Soils that support TDF are usually fertile and attractive to ranchers and pastoralists, leading to widespread clearing of the forests. Certain tree species are also harvested for various purposes, often leaving behind a less diverse habitat. TDF, as it is known to its champions, is especially rich in *Burseras*, leguminous, and euphorbiaceous trees. According to Rzedowski, twenty-one species of *Bursera* occur in the upper Río Papaloapan region, including four endemic species.[58] This total represents nearly one-fourth of the known species of the genus. Some of the *Bursera* species grow exclusively in thornscrub, others only in tropical deciduous forest, while many species are shared by the two vegetation types, and still others find a home in the upper TDF and the oak forests.

The trees of TDF are often interspersed with columnar cacti, which sometimes poke their crowns above the canopy formed by the deciduous trees. In groves protected from heavy overgrazing and woodcutting, even taller trees occasionally tower over the canopy of *Burseras* and columnar cacti. The *pochote*, or kapok (*Ceiba aesculifolia*) and *chupandilla* (*Cyrtocarpa procera*) are especially notable for their size. The morning glory tree, *cazahuate* (*Ipomoea arborescens*), is common in both thornscrub and TDF. It only occasionally exceeds 10 meters in height but stands out

57. We suspect that large tracts of tropical deciduous remain along the upper Río Grande. Inaccessibility, however, has precluded our determining the extent and composition of that forest. The area is roadless, impossibly steep, and inhospitable to the hiker.

58. Rzedowski, Medina-Lémus, and Calderón de Rzedowski 2004, 137.

FIGURE 7.46 Tropical deciduous forest near Jayacatlán, Cañada Chica.

FIGURE 7.47 Slopes of the Sierra Monteflor near Jayacatlán, showing strong slope effect between oak forests and chaparral. Less visible is the upper limit of tropical deciduous forest. Oaks occupy the northern exposures, which retain moisture longer. On the right, pines flourish in the oak forest downslope nearly to the oak-chaparral boundary.

while leafless during the winter months with its numerous large flowers of brilliant white.

In the absence of tropical storms in the fall, which may delay the drying of the forest, the trees and shrubs in tropical deciduous forest typically drop their leaves by November. They then remain leafless until the advent of summer rains in late May or June.

Palm Forest and Scrub

Old Highway 150 connects Tehuacán with Puebla. Roughly 20 kilometers northwest of Tehuacán, a branch turns west from the road and winds upward into the mountains called the Cerro Majada. There it encounters a canyon called Cañada Mora, one of several drainages choked with forests of palms. They form a most unlikely habitat in this semiarid region. Here, beginning at roughly 1,800 meters elevation and extending several hundred meters higher, the low palm *Brahea dulcis* forms a dense ground story—up to fifteen thousand plants per hectare.[59] This plant commu-

59. Valiente-Banuet et al. 2009, 85.

FIGURE 7.48 Low-story oak woodland dominated by the palm *Brahea dulcis*, Santa María Ixcatlán.

nity seems an endless sea of palm branches rising no more than 2–3 meters from the ground in association with small oak trees, which dwarf the palms. The dead palm branches cover the ground and make crossing the landscape on foot a noisy proposition. Huge numbers of *B. dulcis* also grow intermingled with low-growing oaks at around 1,700 meters elevation near Santa María Ixcatlán, where Ixcatec hat weavers take advantage of the proliferation of palms to produce hats sold in the markets of Mexico City. These palm-oak associations are also common on north-facing slopes covering extensive areas on elevations around 2,000 meters in the vicinity of Puerto Mixteco near Tepelmeme and on the drainages emptying on the upper Río Xiquila—a tributary of the Río Salado with outstanding limestone rock formations and crystal-clear blue waters. Another location for viewing *B. dulcis* in habitat is along Highway 135 as it ascends from the cactus forest and dry tropical forest of the lowlands near Cuicatlán, leaving the Río de las Vueltas behind and winding upward into the oak forest. There the palms grow in association with *Beaucarnea stricta, Nolina longifolia,* and *Mitrocereus fulviceps* in a most varied vegetational regime, one we characterize as chaparral complex.

In sharp contrast, on some hillsides of Cerro Majada near Nopala and along watercourses the tall *Brahea calcarea* (also known as *B. nitida*) grows, reaching more

FIGURE 7.49 Grove of palms (*Brahea nitida*) near Santiago Nopala, Valley of Zapotitlán. Though wild, these palms are managed and exploited for their fronds.

FIGURE 7.50 *Palmar* (palm forest) of *Brahea calcarean* near Santiago Nopala, west of Tehuacán.

than 15 meters in height, towering above the oaks and all other trees. These thin palms stand out individually, but far more striking is the forest of thousands of these trees covering the northern slopes in a setting unmatched in the Americas. The palms seem to flourish in an otherwise deeply eroded landscape. They constitute a valuable resource for the dwindling number of palm fiber weavers in the region.

Chaparral Complex

This vegetation group presents the greatest variation in appearance and is the most difficult to describe of all groups in the valleys, yet it forms an important transition between tropical deciduous forest or occasionally thornscrub below and oak or oak-pine forest or woodland above. Chaparral may be absent or nearly absent in places, especially in the steeper grades of the sierras east of the valleys, where TDF grades quickly into oak forest. Yet, it occupies vast acreages in the Mixteca Alta and on the slopes of Cañada Chica and related drainages of the western side of the valleys. In some areas the chaparral consists of dense growth of thick shrubs, scattered small oaks, and other low trees intermixed with palms and a few columnar cacti (typically *Mitrocereus fulviceps* and *Pilosocereus chrysacanthus*), along with an abundance of

FIGURE 7.51 Well-developed chaparral, Cañada Chica. Species more common in tropical deciduous forest also appear in this vegetation.

Agaves, Beaucarneas, Nolinas, and *Dasylirions.* Though the density of the growth in this sort of chaparral (see figures 2.7 and 2.8) makes passage difficult, the plants are seldom thorny, as thornscrub is. In parts of the Mixteca Alta, the ground cover is sparse, making prickly pears, agaves, yuccas, and dasylirions stand out. In other cases, especially in portions of the Mixteca Alta, junipers (especially *Juniperus flaccida*) and hopbush (*Dodonaea viscosa*) abound, appearing to constitute most of the vegetation. A closer look, however, reveals a wealth of plant species ranging from diminutive annuals to large shrubs. Unless the vegetation is rich in *Mitrocereus* or *Beaucarnea,* which invariably stand out, it is often nondescript or, worse, downright boring in its appearance. Calcareous soils are often lacking in fertility and provide nutrients only to those plants especially evolved to deal with their basic soil pH and impervious surface. These well-adapted plants can be numerous, indeed.

Ecologists Patricia Dávila and Olga Herrera-MacBryde, provide a helpful division of chaparral among three varieties:

1. *Sclerophyllous scrub or chaparral.* Many of the shrubs, often growing on limestone beginning around 1,900 meters, have leathery leaves. The mix includes *Arctostaphy-*

los (manzanita), *Ceanothus*, *Cercocarpus* (mountain mahogany), *Condalia*, *Ephedra* (Mormon tea), and small *Quercus* (oaks). Junipers and hopbush (*Dodonaea viscosa*) may also cover large tracts of this vegetation.

2. *Spiny rosette scrub*. Larger shrubs are lacking in this vegetation, usually above 2,000 meters elevation, presenting an open landscape of arid appearance. It is largely composed of a few grasses plus *Agave*, *Hechtia*, *Nolina*, nopales (prickly pear cacti), and yuccas. This vegetation occupies large expanses of the upper valley of Zapotitlán and the Mixteca Alta.
3. *High, scarcely thorny scrub*. This vegetation is rich in *Beaucarnea* spp., *Dasylirion*, *Forestiera*, *Fraxinus*, *Leucaena*, *Nolina* spp., *Ptelea*, and the columnar cacti *Mitrocereus* and *Pilosocereus chrysacanthus*. It is common on the western slopes of Cañada Chica (see figure 3.5).[60]

These distinctions, however, must not be viewed as exhaustive. The nuances visible in the field usually render sharp distinctions as folly and basic classifications as misleading. The chaparral complex, Mediterranean in character and appearance, must be experienced to appreciate its scope and diversity.

Oak Forests and Woodlands Complex

Characterizing vegetation where oaks abound is as challenging as describing the chaparral, even though oaks possess a certain uniformity of leaves and bark. As with *Burseras* and pines, Mexico is prodigal in oak species, the center of the world's oak diversity, with about one hundred sixty species (out of 500 species worldwide, and close to 220 in the Americas).[61] Precise numbers are elusive as a result of both the wild variability in characteristics taxonomists use to describe species and the fact that many oak species hybridize greedily with others, producing offspring that superficially appear to be a separate species. Furthermore, oaks have evolved into a wide array of ecological niches, so that they flourish in habitats from the upper parts of thornscrub through tropical deciduous forest to roughly 3,300 meters in the pine and mixed conifer forests of the higher mountains. Add to these difficulties the frequent renaming of some species as taxonomists pronounce new identities, and the challenges to understanding the oaks of the region become daunting. Oaks may form just a minor component of the vegetation, like the diminutive chaparral species, or grow in nearly pure forests of magnificent trees. Plant explorer Howard Scott Gentry, who spent years studying the plants of the northwestern Sierra Madre,

60. Dávila and Herrera MacBryde 1997.

61. Nixon 2006; Hipp et al. 2020.

referred to the fully developed oak forest as "the pleasant land of the oaks." Indeed, the oak zone usually possesses a warm climate with ample but not excessive rainfall and moderate temperatures.

Oaxaca alone has fifty-two apparent species—what taxonomists call *morphospecies*—of which twenty-four are white oaks, growing mostly at higher elevations, and twenty-eight species of red oak, growing mostly above 1,000 meters elevation throughout the region. Roughly half of Oaxaca's oak species grow in the Sierra Juárez, the bulk of those appearing in the drainage of the Río Grande and the Sierra Monteflor.[62] We presume that some oaks found in the Sierra Zongolica in Puebla are absent from Oaxaca but have no data to support this.

Oaks range in size from sprawling low shrubs in thornscrub and chaparral, such as *Quercus frutex*, scarcely a meter in height with small leathery leaves, to majestic, spreading trees 25 meters tall in oak forests such as the Oaxacan endemic *Q. macdougallii* and several others, trees that are close rivals of the great oaks of New England in size and beauty. They exhibit leaves ranging from long, slender, and delicate to large, round, and tough. Dense forests almost exclusively of large oaks can still be found on the eastern, western, and southern slopes of the Valley of Cuicatlán, generally above 1,500 meters. These fine woodlands, when protected from the chainsaw, harbor many trees of great age and stature and a dazzling variety of leaves. In less protected sites, oaks genetically capable of growing to tree size survive the onslaughts of overgrazing and firewood and lumber harvesting only in lower or scrub form. Some dense oak woodlands, though historically lumbered but now fiercely guarded as communal property, are to be found near Apoala in La Cañada drainage. In this region frequent fog gives rise to dense development of epiphytes, including orchids and bromeliads, which festoon the oaks. These oaks reach 12 meters in height. Especially notable are small, long, dangling strands of *Tillandsia usneoides*—Spanish moss—that form silvery garlands, lending an air of mystery to the oak groves, especially when a layer of fog descends on the forest. Even this forest, however, is heavily scavenged for dead or downed wood for firewood, which poses difficulties for recruitment of juvenile oaks and denies refuge to mammals that might disperse acorns. Other, more isolated stands of oaks are less combed for fuel and thus seem better able to maintain their viability. This vegetation is also clearly visible from a distance, mantling the mountain slopes with a dense green velvety appearance. Such oak zones are often sharply demarcated, making it possible to step from thornscrub, chaparral, pine forest, or tropical deciduous forest into the oak woodland in a matter of a few meters, as in the abovementioned palm-oak communities. (In other areas, smaller oaks form dense growth as well, usually sharing it with other genera of trees.)

62. Valencia Ávalos and Nixon 2004, 221.

FIGURE 7.52 Oak forest near Apoala. Note the bromeliads.

FIGURE 7.53 Oak festooned with Spanish moss (*Tillandsia* sp.), Apoala. (See also figures 7.16 and 7.51).

Botanists are still in the early stages of mapping the species and relationships of oaks in the valleys. In general, however, they have established that oak woodlands occur in the lower reaches of the mountains. Toward the most xeric side, they merge with tropical deciduous forests and thornscrub in sharp ecotones dictated by soil features, specific thresholds of precipitation and temperature, and in the lowermost areas, slope orientation. In the lower elevations, it is common to find oak woodland interspersed with *Bursera* species and *cazahuates* (*Ipomoea* spp.). Such is the case in the neighborhood of Cuicatlán and eastward, up the mountainside to Concepción Pápalo, in the climb from Teotitlán de Flores Magón toward Huautla, and in the abrupt canyon descending from Apoala toward the Cuicatlán Valley. In the upper Mixteca Alta near Tepelmeme and southward, xerophytic oak woodlands include *Quercus castanea* and *Q. rugosa* as the most common species, but also *Q. liebmannii* and *Q. urbani*. These oaks are often found along with pine species like *Pinus leiophylla*, *P. lawsonii*, and in some cases, *P. pseudostrobus*. In the more mesic side of the spectrum, oaks occur in thick, highly diverse forests with a broad array of tall trees. Oak forests merge in a more imperceptible way into the montane cloud forest (*bosque mesófilo de montaña*), mainly in the uppermost reaches looking eastward into the Gulf of Mexico in the Sierra Mazateca, the Sierra Juárez, and in tiny pockets in Sierra Monteflor, places that frequently become enshrouded in fog. The thick, tall oak forests in the upper Río Grande near Ixtlán de Juárez are delightfully diverse. Rzedowski, citing Paray (1951) mentions more than eight large oak species, several pines, and species of *Arbutus, Bejaria, Clethra, Meliosma, Oreopanax, Saurauia, Styrax, Symplocos,* and other representatives of cloud forests. Here, seemingly far away from the dry lands of the valley bottom, but still within the basin, these antediluvian-appearing montane cloud forests prosper.

South of Nochixtlán, highly variable but extensive oak woodlands are present above 2,000 meters. These are rather low and sparse, reflecting a long history of lumbering. *Quercus magnoliifolia* and *Q. urbani* are the most common large-leaved species, while the also common *Q. castanea, Q. conspersa,* and *Q. obtusata* have smaller leaves.[63] In the vicinity of Santa María Ixcatlán *Q. urbani* combines with *Juniperus flaccida* to form a thick, low woodland that often grades quickly into chaparral dominated by the low palm *Brahea dulcis.*[64] *Q. urbani* thrives on calcareous soils and has large, stiff, cup-like leathery leaves. Nearby on volcanic soils, a forest of mixed oaks with great numbers of *Q. peduncularis* combined with other small tree species also harbors considerable numbers of palms. The leaves of this oak are flatter, toothed, and sometimes lobed. At roughly the same elevation across La Cañada above Quiotepec are dense stands of the large-leafed *Q. magnoliifolia.* In the dry season its leaves turn red, then fall, unusual for Mexican oaks.

In the northwesternmost reaches of the Valley of Tehuacán, near Santa Ana Teloxtoc and Santiago Nopala, large stands of the palm *Brahea calcarea* grow in highly calcareous white soils almost devoid of organic matter in areas whose elevations would suggest a proliferation of oak species (see figure 7.50). These palms can reach 10 meters in height. Their height and gray-bluish leaves set them apart from the short *B. dulcis,* which sports brilliant green leaves. Residents of the nearby hamlet El Encinal (The Oak Woodland) remark that the cover of oaks used to be denser. Still, in these palm groves several species of oaks provide the vegetation matrix. These include *Quercus magnoliifolia* and *Q. obtusata* (*encino carrasco*), along with chaparral species like *Cercocarpus fothergilloides* (mountain mahogany) and *Vauquelinia australis.* Other species of this distinct plant association include *Acacia* spp., *Amelanchier denticulata, Brahea dulcis, Buddleia cordata, Citharexylum oleinum, Karwinskia mollis, Nolina longifolia,* and *Yucca periculosa.*

Oak woodlands and forests are complex entities of Holarctic (i.e., temperate Northern Hemisphere) descent. As Rzedowski has pointed out, these associations are often accompanied by a host of species "forming intricate mosaics and complex

63. Rzedowksi 1978, 291.

64. Valiente-Banuet et al. 2009, 88–92.

successional interrelations that often turn up as mixed forests."[65] Oaks intermingle with tropical and subtropical plant communities but have an intimate relationship with conifers, especially pines—another group of plants with boreal affinities. Mexico is a hotspot for both groups, as phylogeneticists who study evolutionary relationships among groups of organisms have determined, so it is not surprising to find their relationships producing all manner of vegetation associations in the cool and sunny tropical mountains of Mexico. As oaks proliferate, so do other plants and their communities, in ways that must be seen to be appreciated. A stroll through the oak forests of western Mexico from southern Sonora through Oaxaca quickly reveals far more than just oaks, particularly in the rich understory, which comprises a bewildering variety of tropical herbs and shrubs. From a strictly aesthetic viewpoint, Mexico's oak forests form the country's most agreeable habitat. We have walked with wonder through Sonora's oak woodlands, which reach their crowning glory in the high foothills of the Sierra Madre Occidental. Hiking among the oaks of Sierra Monteflor filled us with the same feelings of awe.

Coniferous Forests

Pines, scattered or in forests, monospecific (single species) or multispecific, appear in mountains on the east side of the valleys—the Sierra Juárez, the Sierra Mazateca, and the Sierra Zongolica—beginning at about 1,800 meters elevation (at times, though, as low as 1,500 meters) and continuing to the summits at over 3,000 meters. Extensive pine forests also appear in the Mixteca Alta west of La Cañada. While the vegetation of the Mixteca Alta within the boundary of the valleys is primarily chaparral, oak forest/woodland, or scrub growing on caliche, the Mexican government has undertaken a massive planting of pines in the more eroded sections of that region. For example, since 2007 through the *tequio*—mandatory communal work—system, people of Yanhuitlán have reforested 247 hectares (about 610 acres) mainly with pine trees (*Pinus pseudostrobus* and *P. greggii*). Other tree species used for replanting include *Alnus acuminata*, *Quercus acutifolia*, *Amelanchier denticulata*, *Eysenhardtia polystachya*, *Leucaena diversifolia*, *Rhamnus serrata*, *Dodonaea viscosa*, and several *Acacia* species.[66] Similar reforestation programs have been applied throughout the Mixteca with varied success. Although large tracts of land have been reforested, they represent but a tiny amount of the highly degraded upper Oaxacan Mixteca landscape, where about 85 percent of the land shows moderate to high erosion, 13 percent very high erosion, and only 2 percent low erosion.[67]

65. Rzedowski 1978, 295–96.

66. Ibarra Contreras 2013.

67. Narváez Carvajal 1991, 33–40.

FIGURE 7.54 Pine-oak forest, Sierra Cuicatlán, west slope.

FIGURE 7.55 Mixed conifer forest, Sierra Juárez, east slope.

FIGURE 7.56 *Pinus hartwegii* and other high-elevation conifers, Sierra Juárez summit looking east toward the Gulf coast.

Much of the region west of the valleys and at higher elevations is still forested with pines. Stands are often wispy, and most of the region is actively lumbered. Land tenure in much of the Mixteca, however, consists of comunidades. These are communally owned lands in which the trees are viewed as part of the commons not subject to individual appropriation. Harvesting probably exceeds growth, although demands for forest products have slowed in recent years. Most new housing in villages incorporates brick or concrete block into walls and galvanized sheeting for roofs, replacing the traditional pine log sides and shake roof. Demand for new housing has also slackened as hundreds of thousands of Mixtecs have emigrated from the region to northern Mexico and onward to the United States, especially Southern California. They repatriate earnings from the United States into more conventional brick and mortar dwellings in the Mixteca and have thus spared many forests the ax and chainsaw, resulting in lower erosion and deforestation, and perhaps affording in some areas an opportunity for the vegetation to recover.

In the mountains on the east side of the valleys, sawmills are still operating in various locations. Although some forests are protected and managed by Indigenous communities, harvesting appears considerably to exceed growth. The summit of the Sierra Juárez (including Sierra Cuicatlán), east of Concepción Pápalo, burned around the year 2000, leaving large tracts devoid of mature trees, but recruitment of young pines appears to be vigorous. In general, though, Mexican ecologists note the gradual deterioration of the forests due to habitat destruction, overharvesting, global warming, disease, and fire.[68]

The composition of the pine forests varies with elevation, soil quality, slope inclination, and exposure to the sun. The lowest-elevation pines mix freely with oaks, palm trees, occasional columnar cacti, and other trees and shrubs, forming bewilderingly diverse vegetation, as can be seen in figure 7.15. With increased altitude and northern exposure, pines increase in dominance, while oaks increase in size. Pine species appear and fade as they grade across the altitudinal gradient with typical species of low, medium, and high elevation. Mexico Highway 135D 20 kilometers or so north of Oaxaca City passes through oak-pine forest (though now thinned by woodcutters), partially in the headwaters of Cañada Chica and of the Río Tomellín. Equally varied is the ever-changing forest landscape along old Route 135 around the village of El Moral. There the dominant trees are primarily oaks, with occasional pine forests appearing along the roadsides. Fine stands of pines or pine-oak forests are also found in the uppermost reaches of Cañada Chica, readily visible north of San Juan del Estado, where Río de las Vueltas has its headwaters. The Sierra

68. Del Castillo et al. 2004, 141.

FIGURE 7.57 Mixed conifer forest, Sierra Juárez, looking eastward into the Papaloapan drainage, here obscured by cloud bank.

Monteflor, the imposing mountain range that separates Cañada Chica from the Río Grande to the east, supports large forests of diverse pines and oaks. As elevation rises, oaks gradually decrease in proportion as pines increase.

In La Cañada region, common pine species at lower elevations include *Pinus lawsonii*, *P. patula*, *P. pringlei*, and *P. devoniana*.[69] At the highest elevations in the valleys, near the summits of the Sierra Mazateca and the Sierra Juárez at roughly 3,000 meters elevation, pines mix with firs (*Abies hickelii*, *A. religiosa*) in a rich forest heavily dominated by conifers, with fewer species of oaks. Fog is frequent at this elevation, and as a result the trees here are ripe with lichens, bromeliads, and orchids, and the soils are usually moist, a boreal forest deep within the tropics. In these habitats, the shadiest and most humid areas of canyons and ravines, thick growth of firs, interspersed with *Pinus ayacahuite*, *P. pseudostrobus*, and *Cupressus lusitanica*, displace almost any other large arboreal species. In more exposed, drier sites at higher elevation, *Pinus montezumae*, with its rounded canopy, forms extensive forests, and in the highest terrain, in the harshest environments, *P. hartwegii* flourishes in monospecific stands.

---

69. Del Castillo et al. 2004, 152.

Pines of the valley drainages are only recently studied and described in detail, and the taxonomy of many of the species is in flux.[70] Mexico is the world's center of pine diversity, with somewhere around fifty species. The state of Oaxaca is home to at least fourteen species, a number that changes frequently as botanical studies proliferate. Since taxonomy is shifting and pines are as notorious hybridizers as oaks, we note with some hesitation the existence of at least nine species of pines in the valley drainages and suggest that several additional species are probably present, and others will likely be discovered or have already been so. Common pine species include *Pinus ayacahuite*, Mexican white pine, the tallest and most sought after for lumber; *P. devoniana*; *P. hartwegii*, Hartweg's pine, which grows at the highest elevations and is capable of enduring cold temperatures better than any other local pine; *P. lawsonii*; *P. leiophylla*, Chihuahua pine, *tlacocote*; *P. montezumae*, *ocote*; *P. patula*, *pino llorón*, Mexican weeping pine; *P. teocote*, Aztec pine, *ocote*, the most widespread pine tree of Mexico, growing in high, dry, stony sites; and *P. pseudostrobus*.[71] While lumber from some species, especially *P. ayacahuite*, closely related to the white pine (*P. strobiformis*), is highly valued, all pines are harvested in one way or another.

In addition to pines, the montane forests that drain into the valleys are home to two species of *Abies* (fir, *abeto* or *oyamel*) and the Mexican cypress, *Cupressus lusitanica*. At mid-elevations, the majestic sabino, or *ahuehuete* (Montezuma bald cypress), *Taxodium mucronatum* lines permanent drainages in impressive number, making the watercourses visible from afar (see figure 7.14). Two species of juniper (*Juniperus flaccida* and *J. deppeana*) dot the landscape in the arid reaches of the mountains. Common nonconiferous trees include the madrone (*Arbutus xalapensis*), the shrubby manzanita (*Arctostaphylos pungens*), and many oak species like *Quercus acutifolia*, *Q. crassifolia*, *Q. magnoliifolia*, *Q. rugosa*, *Q. castanea*, *Q. peduncularis*, *Q. glaucoides*, and *Q. mexicana*. In these forests thrives a rich flora of broadleaved species, including trees from a host of families with tropical affinities and an understory spectacularly rich with tropical taxa. At higher elevations *Q. laurina* can locally cover extensive areas until the pines suddenly reassert their dominion.

Columnar Cacti, *Burseras*, and Other Unusual Plants of Tehuacán and Cuicatlán Valleys

The valleys harbor a vast array of plant forms and functional types, from minute ephemerals to enormous trees, deciduous and evergreen. Many of the plants, particularly these of the drier and warmer valley bottom, are distinctive dry-land plants

70. Del Castillo et al. 2004, 152.

71. Del Castillo et al. 2004; Reyes Santiago et al. 2004, 18.

with specialized water storage tissues in their leaves, stems, and root systems. Water is stored in what the botanists call the parenchyma, a chlorophyll-free and thin-walled specialized water storage tissue in which each cell is mostly a large vacuole full of water. Water storage often makes these plants distinguishable by their swollen trunks and leaves.

Some trees with a thickened trunk, short height in relation to the trunk diameter, and short branches resembling a bonsai are common inhabitants of the low dry-land valleys. These are described by scientists as sarcocaulescent trees and shrubs. They share the functional group of traits that include smooth fleshy soft-wood stems, often photosynthetic, usually with exfoliating bark that naturally peels away from the trunk and branches. The loss or removal of the old papery bark exposes photosynthetic tissue, while the thick semisucculent stems store water in the woody tissues of the trunk and branches. Sometimes, like the limber bush species of the genus *Jatropha*, the roots are also very important water reservoirs. The champions of the water storage competition are the cacti, in which the whole plant is a water reservoir usually defended by fierce spines. Distinctive groups of species stand out by their prominence, size, and extraordinary peculiarities.

Cacti are a diverse group in which many variations of the water storage strategy occur, from small globose miniatures a few centimeters in diameter and barely visible to gigantic columnar and candelabriform trees. They come in various forms and sizes, but the common denominator for all—except for the most primitive cacti with succulent leaves, resembling spiny trees—are their peculiar adaptations to aridity. Most cacti lack leaves. They also produce photosynthesis in their stems, and to minimize water loss and promote water storage, they maximize their plant volume with the smallest surface area. Cacti can accommodate different opportunities for storing water by expanding and contracting, thanks to their special morphology, which features longitudinal ribs, tubercles, and furrows that allow them to change their diameter without altering other functional features. Their spines, sometimes formidable, serve for defense against herbivores but in some cases for dispersal as well, as seen in the jumping cholla and other species. Spines also assist in thermoregulation, even when reduced to numerous hairy filaments resembling a beard or an old white-haired person, a *viejito*. Cacti also feature a highly specialized photosynthetic system, called crassulacean acid metabolism (CAM), which saves significant amounts of water. CAM plants open their stomata—the tiny pores in their leaves—at night to absorb carbon dioxide when water is less likely to evaporate. The atmospheric carbon dioxide is captured as organic acids, temporarily stored in the cell's vacuoles. The next day these four-carbon acids (usually malate) are reconverted to CO_2, and normal photosynthesis occurs to produce sugars using the light from the sun, with the difference that the stomata remain closed during the day. All CAM plants like cacti, aloes, bromeliads, and many other succulents become progressively acidic through the night, a process scientists recognized long ago.[72]

Cacti probably originated somewhere in northern South America during the Late Eocene or Early Miocene. With the increasing global aridification and lower atmospheric CO_2 of those eras, several dry-land lineages with photosynthetic systems using four-carbon molecules underwent robust diversification. Among these were euphorbs, agaves, and grasses. Not until recently, however, about 2–6 MYA, did these families undergo rapid diversification in both their geographic distribution and their growth forms. In the early evolution of the cacti, several groups split from the main phylogenetic stem. All cacti, however, have specialized structures called areoles, wart-like protrusions, a distinguishing feature of the family. The growing or meristematic tissue of cacti is located within these areoles. Areoles can be seen as greatly shortened branches where a cluster of spines or highly modified vestigial leaves originate, and where new branches, flowers, and fruits develop.

Aside from the early diverging groups, the bulk of the cactus species comprise two major branches. One includes the highly diverse prickly pears, chollas, and their allies. The other encompasses the globose, hemispheric, barrel-shaped, climbing, and large columnar and candelabriform cacti of North and South America. Each of these groups have followed separate paths in North and South America, and their functional forms are the result of original kinship, geographic separation, and convergent evolution.

The valleys of Tehuacán and Cuicatlán are a world hotspot for cactus species, hosting at least twenty-five genera and eighty-nine species.[73] Paraphrasing British evolutionary biologist J. B. S. Haldane, "If a god or divine being had created all living organisms on Earth, then that creator must have an inordinate fondness for beetles" or, in the case of the valleys, for *Mammillaria* cacti. Twenty-six species of *Mammillaria* and nineteen of columnar cacti grow across the valleys (see appendix).

The variety and shapes of the columnar cacti of the valleys are what first attracted us to the area. Columnar cacti are the epitome of the dry-land tropical plants. These often-grotesque growths—some of enormous proportions—not only represent the culmination of adaptations to extreme aridity, but also provide extraordinary examples of biological adaptations in terms of pollination, dispersal, and plant-to-plant interactions. They further provide food and construction materials to human populations of the region. Their massive trunks accumulate more water than any other plant in the world. Their nocturnal flowers are mainly pollinated by bats, but various pollinators, including bees, hummingbirds, moths, hawkmoths, and perching birds

72. Black and Osmond 2005.

73. Arias Montes 2020. See also the Global Biodiversity Information Facility home page, accessed October 27, 2022, https://www.GBIF.org.

TABLE 7.1 Species of columnar cacti growing in the Tehuacán and Cuicatlán Valleys and their common names

Cephalocereus columna-trajani	*viejito* (endemic)
Escontria chiotilla	*jiotilla, xuega*
Isolatocereus dumortieri	*malayo, milayo, candelabro*
Mitrocereus fulviceps	*órgano de cabeza amarilla*
Myrtillocactus geometrizans	garambullo, *padre nuestro*
Myrtillocactus schenckii	*vichisova*, garambullo
Neobuxbaumia macrocephala	*órgano de cabeza roja* (endemic)
Neobuxbaumia mezcalaensis	*órgano, clavija*
Neobuxbaumia tetetzo	*tetecho*
Pachycereus grandis	*cardón, chicomejo* (grows just outside valley boundary)
Pachycereus hollianus	*baboso, acompes* (endemic),
Pachycereus marginatus	*órgano*, malinche
Pachycereus weberi	*cardón, candelabro, tepananochtli*
Pilosocereus chrysacanthus	*viejita*
Polaskia chende	*chende* (endemic)
Polaskia chichipe	*chichipe* (nearly endemic)
Stenocereus pruinosus	*pitayo de mayo*
Stenocereus stellatus	*xoconochtli*

pollinate some species. The delicious fruits, usually red and juicy, are the reward for the dispersal of their seeds. Most columnar cacti require other plants for protection during the germination and establishment phase, which increases the overall plant diversity of the valleys. Eighteen species in ten different genera inhabit the valleys. A list of those thus far documented in the valleys, their common names, and where they grow is shown in table 7.1.

Cephalocereus columna-trajani, the *viejito*, grows abundantly on hillsides immediately southwest of Texcala on Route 125. There the hordes of cacti resemble ranks of soldiers climbing up the mountainsides. Additional populations can be found on hillsides in the southern reaches of the Valley of Tehuacán near Tecomavaca and in mixed thornscrub and tropical deciduous forest on the western slopes of La Cañada on the road connecting Highway 135 with Santa María Ixcatlán and Coixtlahuaca (see figures 7.32 and 7.33). What is perhaps the densest grove appears on the lower slopes of the eastern foothills of the Valley of Tehuacán along the road to San Antonio Cañadas. Scattered individuals can also be found on Cerro Petlanco (see figure 7.34). The *viejito* is easily distinguished from other columnar cacti by its stooped habit, its grayish color, and the bushy pseudocephalium covering much of its north side. It normally grows as a single stem.

Commonly known as *jiotilla* or *xuega*, *Escontria chiotilla*, a candelabriform cactus, is widely distributed in the valleys. Fine stands of this goblet-shaped, yellow-flowered, many-branched cactus can be seen along the *autopista* connecting Tehua-

FIGURE 7.58 *Jiotilla* (*Escontria chiotilla*), Valley of Tehuacán.

cán and Oaxaca where tropical deciduous forest grades into chaparral, 10 kilometers or so south of the Oaxaca state line. It is also abundant in the southern reaches of the Valley of Tehuacán. Its fruits lack spines, sporting scales instead. They are tasty and are harvested and sold in local markets in mid- to late summer.

Though common in some parts of southern Mexico, *Isolatocereus dumortieri,* locally known as *milayo* or *candelabro,* is nowhere abundant in the valleys. Plants can grow to great size and because of their candelabra shape can be confused with the larger *Pachycereus weberi.* Branches have fewer ribs than *P. weberi,* and larger plants often sport branches that, though dead, remain upright, giving the plants a scraggly appearance. Large plants grow above Quiotepec on the winding mountain road to Coyula. In addition, a few plants can be seen in the vicinity of Acatepec, at the southwestern limits of the Valley of Zapotitlán. Some fine specimens grow on hillsides just northwest of Atolotitlán, southeast of Los Reyes Metzontla. Smaller plants grow in pockets of tropical deciduous forest in southern La Cañada and along the *autopista* as it climbs from the Valley of Tehuacán (see figure 7.43).

One of the largest columnar cacti is *Mitrocereus fulviceps.* Its common name, at least in the Zapotitlán Valley, is *órgano de cabeza amarilla,* or *cardón.* It is well represented in pockets in the Valley of Zapotitlán and in Cañada Chica. Second in

FIGURE 7.59 *Xuegas*, fruits of *Escontria chiotilla*.

size only to *Pachycereus weberi*, it penetrates the Cañada Chica well into the oak zone, and from mountain highways is often visible from a considerable distance. Numerous individual plants can also be easily detected rising from chaparral between Santa María Ixcatlán and La Cañada, in an ecological zone above that of *Cephalocereus columna-trajani*, which replaces *Mitrocereus* as the roadway descends toward La Cañada. The branches of *Mitrocereus* grow quickly parallel to the main stem, often equaling or exceeding it in height (see figures 2.7 and 2.8).

On Mexico Route 125 from Tehuacán into the Valley of Zapotitlán, *M. fulviceps* is the first large columnar cactus to appear on the hillsides. While it does not appear to be endemic to the valleys, we have not seen it growing beyond their limits.

Two members of the genus *Myrtillocactus* inhabit the valleys. The more common is *M. geometrizans*, commonly called garambullo or *padre nuestro*. The garambullo is widely distributed throughout central and southern Mexico. Branches are often blue-gray in color and tend to grow enthusiastically from other branches, sometimes by the hundreds. The dark purple fruits grow several to a cluster and are the size of grapes. They are free of spines and hence easily harvestable. In the Valley of Zapotitlán brewers produce a powerful and tasty wine from the fruit and prepare a concentrate from the pulp that is made into a refreshing purple beverage. Beautiful as it is in its native range, in South Africa, introduced *M. geometrizans* plants are invading natural areas. Although the rate of the invasion has been slow, it is likely to

FIGURE 7.60 *Órganos de cabeza roja* (*Neobuxbaumia macrocephala*) near Acatepec, Valley of Zapotitlán.

prove detrimental to many of the endemic succulents of the Karoo if not contained. Ironically, that infestation appears to have originated from a cactus collection that was established about forty ago.[74] From there, insect pollinators and bird mutualists have facilitated its successful fertilization and directed dispersal to several indigenous shrubs that serve as nurse plants.

Myrtillocactus schenckii, commonly called *vichisova*, is rather less common and more narrowly distributed than *M. geometrizans*. The *vichisova* is confined to Puebla and Oaxaca. Its branches tend to grow more vertically and are more yellow-green than those of *M. geometrizans*, and the areoles are much more numerous and closely spaced. Plants are abundant around Los Reyes Metzontla and in the thornscrub and tropical deciduous forest of La Cañada between Cuicatlán and Concepción Pápalo. Some larger plants may reach 8 meters in height, but most are shorter. In the rugged canyon country east of Oaxaca's central valleys, *vichisovas* may reach 10 meters in height (see figure 7.42).

---

74. Dean and Milton 2019.

FIGURE 7.61 Garambullo (*Myrtillocactus geometrizans*), Valley of Tehuacán.

One of the handsomest columnar cacti is *Neobuxbaumia macrocephala*. These giants are known around Zapotitlán Salinas as *órgano de cabeza roja*. They are large shapely plants with several branches emerging from the central column. The spines at the crown or the tips of the branches are usually reddish or pinkish, the source of the common name, and the top meter or so of the branches often appears to be stained dark. The flowers and fruits are confined to the ends of the branches. These plants are narrowly endemic, found only in the southwestern parts of the Valley of Zapotitlán and in the vicinity of San Antonio Cañadas, northeast of Tehuacán in the Sierra Zongolica. Around San Juan Raya they grow in thick forests in combination with *N. mezcalaensis*. Near Los Reyes Metzontla they can be found growing together with *Mitrocereus fulviceps*, a good place to study their differences and similarities. The two also grow sympatrically high on the hills of the Sierra Zapotitlán, north of Zapotitlán Salinas. To see them in this location, however, requires considerable hiking. The hiker may be compensated for the effort by a drink of fresh aguamiel from the plantations of *Agave salmiana* growing near the ridge of the mountains some 400 meters above the town. A fine pure stand of *N. macrocephala* grows on a hillside on the north side of Highway 125 just before arriving in Acatepec, at the southwestern limit of the Valley of Zapotitlán (see figure 7.60).

One of the tallest, if not the tallest, of any cactus is *Neobuxbaumia mezcalaensis*, locally known as *órgano* or *clavija*. These cacti are known to reach 20 meters in height in ideal locations. In our valleys they appear to be confined to the extreme southwestern reaches of the Valley of Zapotitlán where they grow in great numbers, often in the company of *N. macrocephala*. *N. mezcalaensis* branches only when injured. The plants can be difficult to distinguish from *N. tetetzo*, but the latter is usually multibranched, and the flowers grow only at the apex. *N. mezcalaensis* is usually single trunked and arrow straight (see figure 7.37). Its flowers are white and resemble stars, emerging well down the central trunk from the apex. The fruit husks are black.

The *tetecho*, *Neobuxbaumia tetetzo*, may be the commonest columnar cactus in the world, but it grows only in Puebla and Oaxaca (perhaps eastern Guerrero), where the plants flourish in splendid numbers. Dense forests cloak numerous slopes in the Valley of Zapotitlán. In the southeastern foothills of the Valley of Tehuacán, staggering numbers swathe the slopes, tinting the hillsides pale green. In the arid foothills of the Sierra Mixteca, which rise from the western side of the southern valley of Tehuacán, the texture of the mountainsides is softened by the millions of plants that seem to carpet the slopes into infinity. Thick groves appear on foothills well south into Cañada Chica, as far south as Atatlahuca, where they end abruptly. Additional dense populations grow in eastern and southern Oaxaca. The buds, flowers, fruits, and seeds are all considered a delicacy by residents of the Valley of Zapotitlán, where local restaurants serve *pipián*, a sauce made from seeds of the fruits, called *tetechas*, *zaleas*, and sometimes *higos de teteche*. Residents report that a good harvest of fruits occurs only every few years. Since the plants, often more than 10 meters tall, flower only at the apices of the branches, collecting can be a challenge (see also figures 4.1 and 7.35).

The gigantic *Pachycereus grandis* can be found in a large grove near the Cuicatec town of Coyula, on a mountainside far above Quiotepec, overlooking the Río Santo Domingo. Residents there call the plants *chicomejo*. We have not found them in the valleys proper but expect them to be found some day. They are also found in the valley of the Río Acatlán, west of the Valley of Zapotitlán.

The *baboso*—meaning slimy or drooling—cactus *Pachycereus hollianus* (also called *acompes*) is endemic to the valleys, but in the Valley of Cuicatlán it is found only in the extreme northern portion. It is named after the Mexican cactus pioneer Helia Bravo Hollis, whose work we have often used as a basic reference. *P. hollianus* seems to prefer deep soils in valley or arroyo bottoms. It is most abundant in the Valley of Zapotitlán southwest of Zapotitlán Salinas, where it occurs in great numbers. *Babosos* can be unattractive plants, growing in clusters that often sport dead trunks and branches. The fruits are large and sweet and usually grow in pairs at the apices

FIGURE 7.62 *Pachycereus grandis*, Coyula, Río Santo Domingo.

FIGURE 7.63 *Babosos, Pachycereus hollianus*, near San Juan Raya, Valley of Zapotitlán.

of the plants. Harvesting is not difficult since the plants seldom exceed 6 meters in height. The areoles feature a downward facing central spine that often exceeds 8 centimeters in length and presents a formidable defense for the plant. This feature and the ease of propagation makes the plants natural candidates for living corrals that prevent goats, sheep, and cattle from roaming. Those central spines pack a potent wallop for adventuresome livestock seeking to break out of the corral.

The Mexican fence cactus is well named, for it forms thousands of fences throughout southern Mexico. Locally known as malinche or simply *órgano*, *Pachycereus marginatus* plants grow readily from cuttings. Since the cuttings grow straight and do not branch, they can be planted very close together, and the resulting fence will be impossible for goats to pass through. The flowers and fruits are red. The fruits are usually easy to harvest, especially from fences, where they are easily accessible though armed with a few spines. They are often sold in local markets. The plants are seldom seen in the wild, but wild individuals may reach 10 meters in height.

Pachycereus weberi is the world's largest cactus in terms of biomass. It is found in only a few locations in the Mexican states of Guerrero, Oaxaca, and Puebla. The densest stands occur in our valleys, and there they dominate the landscape. The giants grow only at lower elevations of the valleys, below roughly 1,300 meters, and are probably absent from the Valley of Zapotitlán except near the mouth of the arroyo

FIGURE 7.64 *Cardón* (*Pachycereus weberi*), Sierra Cuicatlán.

where it enters the Valley of Tehuacán. They are present throughout La Cañada in impressive numbers. Adjacent to the plaza in Cuicatlán is a mural dedicated to the mighty but defenseless *cardón.*

The widespread common name is simply *cardón,* but in areas where Náhuatl is spoken, the cactus is called *tepananochtli.* The fruits are sweet, so it seems appropriate that the Indigenous name should commemorate the Náhuatl meaning *pleasant fruit* (*tepa* = pleasant, *nochtli* = cactus). While the fruits are delectable, they are often inaccessible because of the size of the plants. In Quiotepec, residents produce a wine from the fruits. Until recently peasants used the massive crotches of the trees to store hay or other fodder to protect it from marauding livestock.

Some columnar cacti sport dense beards of white hairs near the tips of branches. The *viejita, Pilosocereus chrysacanthus,* with its stark white coiffure, is easy to identify. It is popular because its fruits are free of spines. It is nowhere abundant but

FIGURE 7.65 *Viejita* (*Pilosocereus chrysacanthus*), Los Reyes Metzontla, Valley of Zapotitlán.

FIGURE 7.66 *Chende* (*Polaskia chende*) from hills above Los Reyes Metzontla, Valley of Zapotitlán.

is found throughout the valleys, usually on hillsides, where it is easily picked out because of the white clumps on the branches.

The *chende*, *Polaskia chende*, probably has the smallest distribution and the fewest numbers of any columnar cactus in Mexico. In the Valley of Zapotitlán these rare plants are restricted to a small portion of the upper hillsides. They are widely recognized in the area, but wild plants are confined to lofty hillsides. We have found them as well in the high, steep mountainsides overlooking the Río Calapa, which forms the boundary between Oaxaca and Puebla, so they likely grow in restricted habitats in Oaxaca as well. Many yards in Los Reyes Metzontla and in Acatepec have cuttings that have grown into large plants. Accessible large plants grow in abundance along the dirt road connecting Atolotitlán with Caltepec. The fruits are tasty and are also valued for the powerful dye derived from the pulp. *Chendes* seem to flourish as horticulture specialties in the Sonoran Desert of southern Arizona.

A species closely related to the *chende* is *Polaskia chichipe*, the chichipe. Similar in appearance to the *chende*, the chichipe's fruits grow only at the apices of the branches, while those of the *chende* grow well down the branches. Both cacti

FIGURE 7.67 Hillside with a heavy growth of chichipes (*Polaskia chichipe*) in the town of Atolotitlán.

branch profusely, but chichipes are more inclined to branch wildly. Chichipes are also distributed only in the upper reaches of the valley of the Río Zapotitlán and in adjacent valleys and hillsides. Residents have no problem at all in distinguishing the two. A hillside in the town of Atolotitlán has an especially heavy growth of chichipes, forming a landscape labeled by some a *chichipera* (see figure 7.67).

One of the most prolific fruit producers in the region is *Stenocereus pruinosus*. It is the columnar cactus most widely planted for fruit production, which reaches its maximum in May, hence its common name, *pitayo de mayo*. The fruits are usually considered the finest of all cactus fruits. These admirable plants grow readily from cuttings, and thousands now form orchards and produce their tennis ball–sized fruits in May and often in October as well. Many are harvested, packed (with great care), and shipped to Mexico City, where they quickly sell out. In rural portions of the valleys, yards of many, if not most, homes have one or two of these cacti as a source of free fruits. The trunks and branches have seven ribs or fewer, which helps distinguish *S. pruinosus* from other members of the genus *Stenocereus*. The plants may reach 10 meters in height. The flowers are white, and flowers and fruits descend well down the branches from the apices.

Nearly as popular as the *pitayo de mayo* is the *xoconochtli, Stenocereus stellatus.* It is also widely planted as a horticultural tree and as part of living fences. Its fruits come in a variety of colors, each with its own flavor, and are somewhat tarter than those of *S. pruinosus.* The plants are much shorter as well, seldom more than 4 meters tall. The ribs number eight to twelve. The flowers are pink and grow exclusively in a circle on the apices of the branches. They are less inclined than *S. pruinosus* to branch when young, and if cuttings are planted close together in a line, they will form a productive living fence. *Xoconochtlis* produce fruit in the fall, mostly October, which leads to their second common name, *pitayo de octubre.* Closely related to the *xoconochtli* is the *tunillo* (*S. treleasei*), a shorter cactus that produces multiple blossoms and fruits at its apex. We have not found it in the valleys.

Agaves (Agavaceae)

On par with the richness of columnar cacti, *Agaves* (Agavaceae, within Asparagaceae) and their allies are distinctive elements of the valleys' landscape. Unbeknownst to most people, they belong, in a broad sense, to the asparagus family. They range from minute plants, not larger than the palm of the hand, to gigantic species covering a circle 4 meters wide. Agaves are known for their peculiar morphology of shortened succulent stems, fleshy leaves, and xerophytic habit, typically forming rosettes. Their leaves are usually heavily defended by jagged, curved marginal spines, and a sharp terminal spine. Agaves are perennial species differing from most other plants because each rosette flowers once and then dies. Botanists call this habit a monocarpic rosette. When they reproduce, they offer one of the most spectacular floral displays in nature, after allocating all the energy stored for many years to the production of a large spike—up to 12 meters tall—with many flowers, in some cases many hundreds of flowers. The genus has been traditionally divided in three major subgenera: the *Littaea,* typically with inflorescences arranged in spikes; the *Agave,* properly with branched inflorescences bearing clusters of flowers; and the *Manfreda* subgenus, which comprises several genera with rosettes that flower many times throughout their life. Recent work by geneticists, however, has shown that these groupings are not well supported, as some species recognized as *Agave*—the striata group—are more closely related to *Manfreda* and associated genera than to other species now recognized as *Agave* sensu stricto. Agave afficionados are having to learn a multitude of new taxonomic labels.

Many *Agave* species produce runners. Other plants extend suckers from the stem that replace the rosette after reproduction. Still others produce "pups"—young plants, also called bulbils—that cover the inflorescence where unpollinated flowers

FIGURE 7.68 *Xoconochtli* (*Stenocereus stellatus*; *left*) and *pitayo de mayo* (*Stenocereus pruinosus*; *right*), San José Mogote, Oaxaca.

were located. This phenomenon only occurs when pollination fails to produce fruits; otherwise, fruits produce hundreds of seeds from capsules with three partitions.

In evolutionary terms, the dates of origin and diversification of this diverse and complex group occurred during the Late Miocene about 6.2 million years ago, when the arid biomes of North America were established. With forests disappearing and being replaced by more arid savannas and grasslands, the rate of speciation increased, reaching a peak during the Late Pliocene, about 2.7 million years ago. In the early diverging groups, compact inflorescences forming spikes were the prevalent type. Later, the inflorescence diverged toward more branched inflorescences, forming what botanists call a panicle.[75]

Mexico is the world's agave center, with between 110 and 150 species of *Agaves* sensu lato, recognized to date, more than 75 percent of all known members of the genus.[76] Agaves grow throughout the country, often in abundance, and several species are widely cultivated. Wild varieties are especially well represented in the valleys, which are home to Mexico's highest concentration of *Agave* species. We list only a smattering of the roughly thirty-five species that populate the valleys. Some of the names are tentative. Though much work has been done describing and distinguishing members of the genus, much work remains. We await emerging pronouncements on *Agave* taxonomy.

Agaves are commonly referred to simply as *magueyes*, but several species have their own common names, a reflection of their variety of uses and importance to native peoples. They rank as high as cacti in their emblematic role in Mesoamerican cultural history, where for millennia they have provided fiber, food, tools, industrial materials, and, above all, strong drink. *Agave angustifolia*, known in Oaxaca as *espadín* (little sword), is a pan-Mexican species, and its close relatives (including *A. tequilana*) are widely planted. They form industrial crops in Jalisco for producing tequila (the domesticated *A. tequilana*) and in Oaxaca (the semidomesticated *espadín*) for commercial production of mezcal. In the valleys, mezcal is readily available in Oaxaca, a common mealtime accompaniment in Teotitlán de Flores Magón, not far from the Puebla state line. Across the line in Puebla, however, including the Valley of Tehuacán, mezcal is absent or suppressed. In affluent cultures of some states north of Oaxaca, mezcal is viewed, along with pulque, as a drink of the lower classes. Those fatuous snobs do not know what they are missing.

Agave americana. This cosmopolitan agave has several subspecies, appearing in a variety of habitats. It is found at most elevations in Mexico. It is also grown as an ornamental plant in subtropical and mild-temperate locations throughout the

75. Jiménez-Barron et al. 2020.

76. García-Mendoza 2004, 161; Delgado-Lemus et al. 2014. See also Gentry 1982.

FIGURE 7.69 *Agaves* of Cuicatlán and Tehuacán Valleys: (A) *Agave americana*, San Pedro Nodón; (B) *Agave mapisaga*, San Antonio Cañada; (C) *Agave salmiana*, Zapotitlán Salinas; (D) *Agave salmiana*, San Bernardino Laguna; (E) *Agave rhodacantha*, Tepelmeme; (F) *Agave marmorata*, Cerro Cuthá; (G) *Agave potatorum*, El Parián; (H) *Agave* aff. *cupreata*, Acatepec.

FIGURE 7.70 Agavaceae of Cuicatlán and Tehuacán Valleys: (A) *Agave* aff. *triangularis*, Apoala; (B) *Agave stricta*, Tepelmeme; (C) *Agave seemanniana*, Quiotepec; (D) *Agave quiotepecensis*, Quiotepec; (E) *Agave kerchovei*, San Juan Raya; (F) *Agave nussaviorum*, Tepelmeme; (G) *Agave macroacantha*, Cerro Petlanco; (H) *Furcraea macdougalii*, Los Reyes Metzontla.

FIGURE 7.71 *Agaves* of Cuicatlán and Tehuacán Valleys: (A) *Agave chiapensis*, Apoala; (B) *Agave karwinskii* (and *Stenocereus treleasei*), Ocotlán, Oaxaca; (C) *cacayas*, edible buds of *Agave rhodacantha*, Nochixtlán; (D) *Agave kerchovei*, Apoala; (E) *Agave* aff. *cupreata*, Ixcatlán; (F) *Agave* sp. in bloom, Tomellín Canyon; (G) *Agave guiengola*, Sierra Guiengola, Oaxaca.

world. It has become naturalized in the West Indies, South America, the Mediterranean coast, Africa, Asia, and Australia. The variety pictured in figure 7.69 may be subspecies *oaxacensis*. It can grow to be enormous, one of the two largest we have seen, and is well suited for production of pulque and mezcal. It is impractical for pulque, however, since the plants require a couple of decades to mature. It is frequently planted in yards, fences, and waste places to hold soils or terraces in place, in addition to its virtue of yielding precious beverages, as well as its ornamental beauty.

Agave rhodacantha. We suspect that this large plant growing near Tepelmeme was cultivated for fiber production, for which it is known in nearby Nochixtlán and elsewhere. It sports straight spear-like leaves of a blue tint. The flower buds, called *cacayas*, as with many agave blossoms, are edible and marketed in Oaxaca and Puebla. The taxonomy of this species is confusing, for it shows remarkable affinities with agaves growing far to the north.

Many agave species prefer the habit of clinging to rock walls, perhaps as an escape from herbivores. The same applies to many species of bromeliads, particularly in the *Hechtia* genus. *Agave chiapensis* frequents canyon walls, especially limestone, where it can often be found on sheer cliff faces, as we found in Apoala. Near the Río Hondo and the Calapa Bridge, on limestone cliffs, *Agave oteroi*, a fiercely armed agave with a large woody margin below the terminal spine, was described in 2019 by Starr and Davis. *Agave guiengola* is another cliff-loving species. We have not seen it in the valleys, but we mention it here because it is native to Cerro Guiengola, a limestone mountain in southern Oaxaca in what was the last enclave of the Zapotecs. It is a highly distinctive agave with broad, pale, bluish leaves that have densely serrated margins. It grows on cliff faces in great numbers. The famous agaveologist Howard Scott Gentry mentioned that this species resembled dense colonies of sea-stars.

Agave mapisaga. Agave mapisaga grows around 2,000 meters elevation in the Sierra Zongolica and in Oaxaca as well. Plants become gigantic and are harvested for pulque and, apparently, for mezcal. It is not clear whether they occur in the wild or are strictly a domesticated cultivar, probably the latter. They produce fewer offshoots than *A. salmiana* so are less favored for production of pulque.

Agave karwinskii. This is the only agave in the region that grows a trunk, in this case a distinct trunk—hence the local name *barril* (barrel). It is common in the valleys, often in association with *Neobuxbaumia tetetzo*. It is widely planted for delimiting fields. In a few locales, it is harvested for producing *cuishe*, a distinct mezcal with a biting flavor. When allowed to grow taller before being harvested, is called *madrecuishe*. Other vernacular names are *cirial* (candle), *tripón* (large gut), and *tobasiche*.

Agave kerchovei. This bright green agave is medium sized and grows throughout the valleys at lower and mid-elevations. It is a valuable source of ixtle, fiber used for producing rope, mats, and other artifacts, and of edible *cacayas*, flower buds that are commonly eaten. It is also the source of the thick, full-bodied mezcal *jabalí*, also called *mexicano*. Mezcaleros report that producing mezcal from this species requires special attention because of its tendency to produce foam when placed in water after roasting. Agave specialists have devoted considerable efforts at clarifying this species and its close relatives. It is closely related to the narrowly distributed *A. titanota*, *A. oteroi*, and *A. quiotepecensis*. In this group Gentry lumped *A. convallis* with *A. kerchovei*, but Abisaí García-Mendoza considers it a distinct species.[77] The scientific name is subject to change.[78]

Agave macroacantha. Commonly called *maguey de cacayas* for the delectable edible flowers that are commonly sold in regional markets, these smallish handsome plants are found throughout the valleys and commonly mark property boundaries in rural areas. The leaves are rather succulent but taper quickly to a sharp point.

Agave marmorata. *Agave marmorata*, called *pisomel*, grows throughout the valleys. It is especially common on Cerro Cuthá. *Pisomel* has large broad leaves, bright green to bluish in color. It is the source of the mezcal called *tepeztate*.

Agave potatorum. *Agave potatorum* forms a complex of species. It is commonly known as *maguey tobalá*. *A. potatorum* is a medium-sized agave with wide leaves and zigzag margins. It is especially common on rock faces along the highway in the oak zone of Mexico 135, south of Santiago Dominguillo. This is the plant from which the renowned mezcal *tobalá* is distilled, one of more than twenty varieties distilled from wild agaves. Of all wild agaves, it is the most overharvested for mezcal production. *Agave cupreata*, from which a prized mezcal is also produced, resembles a large *tobalá* with even wider, sometimes reddish leaves. This agave is commonly known as *papalometl* (*papalotl* = butterfly and *metl* = agave in Náhuatl), although considerable confusion on the identification occurs among producers.

Agave quiotepecensis. *Agave quiotepecensis*, sometimes called *maguey de pasmo* or *rabo de león* (lion's tail), or perhaps a domesticated descendant is unusually common on the archaeological site of Cerro Quiotepec. We believe it was cultivated there by Zapotec occupiers two thousand years ago. It seems to have a narrow distribution, occurring only in and around the area where the two valleys merge. Its leaves are widely believed to have curative properties. It may be used in mezcal production, but we have yet to verify such a noble application. It is closely related to *A. oteroi* and *A. titanota*.

Agave salmiana. This is the *maguey pulquero* (pulque agave), found growing at around 2,000 meters in fencerows, near homes, in pastures, and in remote planta-

77. Starr and Davis 2019.

78. Starr 2012, 268.

tions. It was domesticated many hundreds of years ago and is considered a valuable resource wherever pulque is drunk. It is probably the most common agave to be found in the oak and pine zone, where it is routinely planted for future pulque production. Wild populations may coexist with domestic plants. Excellent plantations can be found near El Moral, Oaxaca, and near San Bernardino, Puebla, in the Sierra Zongolica above San Antonio Cañadas in the Valley of Tehuacán. It reaches 2 meters in height and width, with huge leaves. Plants are often claimed by *pulqueros* (pulque producers) as their property.

Agave stricta. Agave stricta is a basketball-sized agave commonly known as *erizo* or *gallinitas.* Its flower buds, known as *cacayitas,* are edible, often served with scrambled eggs. Its leaves are very narrow, and the plants resemble large prickly balls. Thick stands cover hillsides east of Puerto Mixteco, a hamlet several kilometers north of Tepelmeme, in the canyon where the natural bridge called Puente Colosal is located.

Agave titanota. Agave titanota is another member of this group, known as *Marginatae* by botanists. It has a very narrow distribution along the Río Xiquila and grows abundantly on the limestone cliffs near La Huerta, which is its center of distribution.

Additional species of agave are common in the valleys, including *A. triangularis, A. nussaviorum, A. seemanniana* (*titanota?*), and *A. convallis,* among others. All have similar uses and local names. We predict that new species are waiting to be discovered, especially in the Sierra Monteflor. Others are merely waiting to be renamed.

Furcraea

The arborescent genus *Furcraea* is closely related to *Agave.* Two species grow in the region and bear mentioning. They are the endemic *tehuitzote* (*F. longaeva*) and *F. macdougalii. Furcraeas* are among the most spectacular plants of the agave family. At the time of reproduction, they attain heights up to 9 meters, including a stem that may reach 6 meters in length. More commonly, the plants are 3–6 meters tall. Reproduction often occurs only twenty to fifty years after germination. Because of their often-gigantic proportions, they are grown as ornamentals near the edges of fields. They rival agaves in providing ixtle fiber for cordage and provide a convenient soapy foam for washing bodies and clothes when the leaves are crushed in water.

Yuccas

Yuccas and their allies are perennial shrubs and trees within the asparagus family. They form rosettes of sword-shaped leaves, in some cases with rigid, sharp terminal spines and serrated edges. During the flowering season, they produce large

FIGURE 7.72 *Furcraea macdougalii*, Chicagua, Oaxaca.

FIGURE 7.73 Izotes (*Yucca periculosa*), Texcala.

terminal inflorescences of white, greenish, or whitish flowers. They can be found from southern Canada to Guatemala, but their highest diversity is in Mexico. They are generally found in arid to semiarid regions, from deserts and coastal dunes to woodlands.

Yuccas are known for their highly specialized pollination system, featuring the small yucca moths (*Prodoxidae*) that lay an egg in the flowers they pollinate. The larva feeds on the developing seeds but leaves enough seeds to make the relationship worthwhile from the plant's point of view. This pollination mutualism is a textbook example of what researchers call *coevolution*. Yet, the interaction is not always as smooth as predicted. The outcome ranges from a relationship in which plant and moth each benefit, called *obligate mutualism*, to one in which one of the partners receives no benefit (ecologists call this interaction *commensalism*), to *antagonism*, where yucca moths do not assist pollination but continue to lay eggs and use the plant resources.

The Yuccas and related genera, such as *Beaucarnea*, *Dasylirion*, and *Nolina*, are prominent species throughout the valleys, in some cases forming dense stands of trees or thickets of shrub yucca and yucca-related species. *Yucca periculosa*, the *izote*, is an abundant, often towering yucca of central Mexico, frequently growing in clusters or small groves. It is common and unmistakable in thornscrub into

FIGURE 7.74 *Yucca mixtecana*, Nacaltepec, Cañada Chica

chaparral in the Valley of Tehuacán, presenting a characteristic appearance of a plant of Jurassic times (though it is not). It reaches 10 meters in height and lives to a great age.

A more solitary yucca is *Yucca mixtecana.* It is far less common than *Y. periculosa,* mostly confined to the valleys of northern Oaxaca and their environs. It also has sparser foliage and appears to be slimmer.

Nolinaceae (Bear Grasses, Ponytails)

Dasylirion serratifolium. Dasylirion serratifolium, a relative of the agave, is common in La Cañada and its tributaries as well as in the Mixteca Alta. It flourishes in lower elevation chaparral. The leaves and bases are gathered and woven into adornments for churches during fiestas. A closely related species in the Chihuahuan Desert, *D. wheeleri,* is the base for the potent distillate *sotol.*

Beaucarnea gracilis (*sotolín, pata de elefante*). Although this mostly endemic plant with its huge tapered trunk can be found growing in the northern portion of the Valley of Tehuacán, it is best viewed in the vicinity of Zapotitlán Salinas, where it stands out so obviously that it is hard to avoid seeing it.[79] Around San Juan Raya it forms fantastical woodlands. Native observers report that young plants are seldom found, indicating difficulty in recruitment, although they are easily grown in horticultural settings. We have not found it outside the Valley of Tehuacán (see figure 7.39).

Two additional species of *Beaucarnea,* or perhaps one and the same species, are *B. purpusii* and *B. stricta.* These species prefer higher elevations than *B. gracilis,* including thick chaparral. If it is a distinct species, *B. stricta* is apparently endemic to the Río de las Vueltas region of the Valley of Cuicatlán. These plants are somewhat slimmer and slightly taller plants than *B. gracilis,* reaching heights of 10 meters, but still with a clearly swollen trunk. They are best seen during the dry season, when their cream-colored flowering stalks are clearly visible among the brown of the dry chaparral (see figure 2.7). *B. purpusii* is abundant east of Xochiltepec on the road from Reyes Metzontla to Atolotitlán.

The bear grass *Nolina longifolia* is a graceful plant abundant in chaparral, especially on the sides of Cañada Chica. The plants are dioecious—having separate male and female sexes. Older individuals form a trunk that may reach 3 meters in height. The long, narrow leaves grow from the crown, then droop like a huge green toupee. The related *N. parviflora* is more treelike, with shorter, straighter, and more

79. Cardel et al. 1997, 367.

FIGURE 7.75 Sotol (*Dasylirion serratifolium*), Cañada Chica.

FIGURE 7.76 Burro packing leaves of sotol (*Dasylirion serratifolium*) to adorn the church of Jayacatlán, Cañada Chica.

pointed leaves. Taxonomists disagree as to whether we are dealing with two species or one with considerable variation. This genus is named after the arboriculturist Abbé Nolin, director of a royal nursery in eighteenth-century Paris that specialized in foreign plants.

Fouquieriaceae

If the cacti and the agaves are unusual and odd-looking plants to those accustomed to the vegetation of northern climes, the Fouquieriaceae—the torchwood, or ocotillo, family—are particularly strange. They have their own small family of only eleven species with only one genus. The family's distribution is confined to what the eminent Mexican botanist Jerzy Rzedowski called Megamexico. The genus includes the ocotillo (*Fouquieria splendens*), the only species found within the United States, where it grows abundantly in the warmer deserts of the Southwest, and the grotesque boojum tree, or *cirio*, of the Baja California peninsula (*F. columnaris*). Members of the genus assume forms that range from spiky, elongated woody stems to treelike to fat, pointed, parsnip-like plants with many radiating small branches. The spines on branches are the hardened petiole of the very first leaves produced on the branch. The leaves themselves shrivel with the arrival of the dry season or

FIGURE 7.77 *Nolina longifolia*, Nacaltepec.

FIGURE 7.78 *Nolina parviflora*, San Antonio Cañadas.

FIGURE 7.79 *Fouquieria formosa* and *F. purpusii*, Cerro Petlanco.

FIGURE 7.80 *Tlapacón* (*Fouquieria formosa*), Valley of Tehuacán.

drought, but the pointed petiole remains for years. New leaves on old branches are sessile, meaning they attach directly to the branches without a petiole. The closest related plant family, or sister family, is the Polemoniaceae, a largely North American and predominantly northern temperate family. Herbs, including many garden cultivars of the genus *Phlox*, are prominent in the family, which also includes a few shrubs and lianas.

Fouquieria formosa. *Fouquieria formosa* (*tlapacón*) grows in thornscrub and tropical deciduous forest throughout the valleys. During the dry season it presents a yellow-brown aspect and a somewhat swollen trunk with spiny octopus's arms, curly spiky sticks emerging at all angles. It leafs out quickly following rain and produces prominent reddish blooms. The plants may reach 10 meters in height. A similar species, *F. ochoterenae*, can be expected in the valleys somewhere near Acatepec, southwest of Zapotitlán. It is known to grow in the adjacent Acatlán valley in Oaxaca and Puebla, which drains into the Río Balsas and the Pacific Ocean.

Fouquieria purpusii. Fouquieria purpusii (*barrilito, jarilla*) is an extraordinary endemic plant, highly restricted in distribution. It grows in a small area in northwestern Oaxaca but is to be looked for in adjacent southern Puebla as well. Folks in the village of Santiago Coatepec, Puebla, near the Río Calapa and Oaxaca, include it in the list of plants that grow in their communal lands. Eighteen plants grow on Cerro Petlanco, situated on or near the state line, and on semiarid limestone hillsides in northern Oaxaca. The plants, which reach a height of more than 3 meters, resemble nothing more than an inverted parsnip (see figure 8.3).

Zamiaceae

Mexico is a hotspot for many plant and animal species. Few are as ancient and curiously individual in their makeup as cycads, which first appear in the fossil record during the Permian period, more than 280 million years ago, well before the dinosaurs.[80] Mexico hosts forty-five species of cycads sensu stricto within three genera, *Ceratozamia*, *Zamia*, and *Dioon*.

Cycads of the genus *Dioon* are most unusual and attractive plants, especially for collectors, which poses a problem for wild populations. They are ancient cone bearers, relatives of pines, little changed from their origins during the Early Paleozoic. Cycads are dioecious, females producing large cones every few years. They are also slow growers, reproducing only at considerable age, and are especially long lived. They resemble both palms and ferns, and though the plants are nearly everywhere protected by law, plant scavengers seem always to find a market for wild plants excavated illegally from their native habitat. In the valleys they usually grow festooned with epiphytes and, often, with multiple *Mammillaria* cacti as epiphytes. The leaves and bases are frequently harvested by residents to adorn churches and plazas during festivals.

The valleys harbor four species of the genus (often referred to in Mexico as *palma real* or *palma de la virgen*): *Dioon purpusii* in Cañada Chica, *D. califanoi* in the Sierra Mazateca, *D. caputoi* in Sierra Mixteca (and near Santiago Coatepec), and *D. rzedowski*, a close ally of *D. spinulosum* in the canyon of the Río Santo Domingo, east of Quiotepec, outside the valleys but close to their edge.

All four species are restricted to small, fragmented, and highly vulnerable populations where overcollecting has further imperiled their survival. Poaching of cycads

80. Locations of some cycads remain a carefully guarded secret among plant researchers. Poaching of the plants is an ongoing problem and equals habitat destruction in endangering the perpetuation of some *Dioon* species.

FIGURE 7.81 *Dioon purpusii*, Cañada Chica.

FIGURE 7.82 *Dioon califanoi*, Sierra Mazateca.

continues at an alarming rate, not solely by foreign collectors, but by local horticulturalists and nurseries as well. In searching for the plants, we inquired about them in a village in the Valley of Tehuacán, describing them and asking if anyone had seen them. One fellow responded that he and his brother collected them from a site in the foothills and marketed them in the city of Tehuacán.

Dioon purpusii. D. purpusii has gray-green lance-shaped leaflets on a flat plane. It is reported from a limited area in the states of Oaxaca and Puebla. The plants often grow on steep hillsides in dense low forests or chaparral, which provides concealment and at least some protection from pillaging. The oldest specimens can attain up to 5 meters (16 ft.) in height and achieve an age of several hundred years. The species name honors the famous plant collector Carl Albert Purpus of the early 1900s, after whom a host of plant species were named, including the rare *Fouquieria purpusii*, the columnar cactus *Pilosocereus purpusii* from coastal Oaxaca, and the elegant *Beaucarnea purpusii.*

Dioon califanoi. D. califanoi was named after Luigi Califano, a former director of the Naples Botanical Garden (Orto Botanico de Napoli). It has strongly keeled (V-shaped) smooth (nonspinulose) leaves. Old specimens can reach up to 3 meters (10 ft.) in height and often carry odd epiphytes, such as the diminutive cactus *Mammillaria flavicentra* and succulent *Echeveria nodulosa,* in their trunks. We found them growing on very steep slopes in the Sierra Mazateca, far above the floor of the Valley of Tehuacán.

FIGURE 7.83 *Dioon caputoi*, Coatepec, Puebla. Photo by Jeff Chemnick.

According to Paulo de Luca and colleagues, *Dioon caputoi* (named after Giuseppe Caputo of the Naples Botanical Garden) is restricted to two small groups of not more than fifty specimens scattered in a remarkably steep open area of about one square kilometer in the ecotone between tropical dry and oak forest in the Río Calapa drainage in the Sierra Mixteca. Jeff Chemnick reports that this population is guarded from plunderers by the adjoining communities. We were guided to the spot by one such guardian.[81]

Dioon rzedowksi. D. rzedowksi, although closely related to *D. spinulosum*, differs by the unequally tapering of its leaflets toward the apex, its entire margin, and the disappearance of spinose leaflets in adult plants. These two species are the tallest of American cycads, with very large leaves, up to 2 meters long. *D. rzedowski* grows on limestone cliffs in tropical dry forest, while *D. spinulosum* is more a plant of truly tropical wet forest conditions, from Oaxaca to Veracruz. It was widely cultivated by wealthy hacendados in the henequén plantations of Yucatán. It has not been reported from the valleys, but along the nearly inaccessible drier lowland forests of the Río Santo Domingo, east of Quiotepec, it occurs in local populations.

The *Burseras*

Finally, we consider in further detail one of the Americas' most attractive and unusual plant genera, the *Burseras*. The valleys of Tehuacán and Cuicatlán have splendid examples of plant associations found only in the region, but the *cuajiotal* is among the most intriguing and visually attractive. *Cuajiotales* are plant communities dominated by *Bursera* species, called *copales* and *cuajiotes* in southern Mexico and *torotes* in the northern Pacific coastal region. These are curious little to mid-sized trees often growing into fat, burly trunks, with gnarled branches, producing strongly scented resins. The genus is part of the Burseraceae, a small plant family with about 540 species in seventeen to eighteen genera, which comprises trees and shrubs native to tropical Africa, Asia, and the Americas. Although the family is thought to have originated in North America—most likely in what is now Mexico—its largest diversity is now found in the Southern Hemisphere, especially in Malaysia, Africa, and South America. It includes the copal, the ceremonial incense burned atop the pyramids of Mesoamerica and the biblical frankincense and myrrh of the Arabian Peninsula and the Horn of Africa.

The resins of most of the Burseraceae are relatively simple terpenes and terpenoids—also known as isoprenoids—which are major building blocks of almost every organism. These essential plant oils are the famous *quintessential* of Paracel-

81. Chemnick 2013, 19ff.

sus, the Renaissance scholar who founded the science of toxicology. The resins of the genus *Bursera* of Mexico, and of *Boswellia* and *Commiphora* of Africa, have been used since antiquity in religious ceremonies throughout the world, and they have also been employed in the production of perfumes, incense, and medicine. Copal, the resin of several species of *Bursera*, typically those with smooth bark, is a staple of native markets throughout Mesoamerica. A wide variety of resins can be found in markets in the Cuicatlán and Tehuacán Valleys as well.

Eighty-nine Mexican *Bursera* species have been described, and about twenty more still await description.[82] Of these, eighty are within Mexico's boundaries, thirty-eight are known from only two states, and nineteen are narrow endemics in single localities. Seventy-nine of these species prosper in tropical deciduous forest.[83] Twenty-one species, about one-fifth of the known Mexican species, occur in the Tehuacán-Cuicatlán region, and twenty-three species are known only from the Balsas River Depression, which drains to the west into the Pacific Ocean. Only two species occur in the United States—the gumbo-limbo (*Bursera simaruba*) of the Florida Everglades and the elephant tree (*B. microphylla*) of extreme southwestern Arizona and southeastern California. The affinities to arid lands are evident by the repeated shifts to xerophytic scrublands (nine species), to oak forests (seven species), and to tropical forests.[84]

Bursera flowers, although varied in shape, are usually very small and, at first glance, nondescript. On inspection with a magnifying lens, however, they reveal a startling variety of colors, shapes, and configurations. Identification of the species is often simplified by the dramatically different patterns, shapes, and sizes of leaves, as illustrated in drawings of leaves of the more common *Burseras* of Tehuacán and Cuicatlán Valleys (see figures 7.89, 7.90, and 7.91).

Although widespread across the valleys, *cuajiotales* are best seen in the southeastern portions, where they form fine stands, especially on the lower eastern slopes of La Cañada and in the vicinity of Cerro Petlanco. The species include *Bursera glabrifolia*, growing with the *tetechos*, *B. morelensis*, *B. aptera*, and *B. submoniliformis*, common in the savannas of the southwest valley of Tehuacán; *B. cinerea*, related to the gumbo-limbo of the Florida Everglades, along the more mesic canyons in tropical deciduous forest; and *B. biflora*, *B. fagaroides*, *B. galeottiana*, *B. linanoe*, and *B. schlechtendalii* on rocky slopes. An unusual *Bursera*-dominated savanna covers several square kilometers south of the Cerro Petlanco, which we discuss below. *Bursera* is a fascinating, attractive, and ancient genus with a center of diversity in Mexico.

82. Rzedowski and Calderón de Rzedowski 2013.

83. Becerra, Noge, and Venable 2009.

84. Rzedowski, Medina-Lémus, and Calderón de Rzedowski 2004; De-Nova et al. 2012.

FIGURE 7.84 *Bursera simaruba* and *Agave guiengola*, Cerro Guiengola, eastern Oaxaca.

FIGURE 7.85 *Bursera* savanna, Cerro Petlanco.

Adding to their strange life form of thickened trunks and branches, some *Bursera* species have exfoliating bark in various colors: silver, red, or yellow. These are locally referred to as *cuajiotes*. Beneath the exfoliation, the bark is greenish, the shade of green depending on the species. The other group of *Bursera* species (mostly section *Bullockia*), locally known as *copales*, mostly have nonexfoliating, smooth and gray to grayish purple bark, as well as highly aromatic resins, which for centuries have been gathered and burned as incense in religious ceremonies. The sister group of *Bursera*, the genus *Commiphora* of northern Africa and Arabia and, perhaps, South America, is the source of the biblical myrrh.[85]

Bursera's family origins have been dated to the upper Cretaceous by Becerra (ca. 93 MYA), and more recently by Weeks et al. (ca. 91 MYA).[86] Some authors support a younger age, for example, Weeks et al. indicated that the divergence between Anacardiaceae and Burseraceae occurred about 60 MYA, and De-Nova and colleagues dated crown-group Burseraceae to the Early Paleocene, about 65 MYA, or earlier.[87] Although a Gondwana vicariance—the separation of America and Africa long ago—has long been the favored explanation of the origin of the genus, a more recent

85. Becerra et al. 2012.

86. Becerra et al. 2012; Weeks et al. 2014; Weeks, Daly, and Simpson 2005.

87. De-Nova et al. 2012.

FIGURE 7.86 *Bursera cinerea*, Cañada Chica.

FIGURE 7.87 *Bursera morelensis*, Cerro Petlanco.

FIGURE 7.88 Squirt defense, *Bursera schlechtendalii*, Cerro Petlanco.

boreotropical dispersal seems to better explain an American origin of the genus during the Eocene, about 50 MYA. The evolution and species radiation through time has shaped a complex biogeography, which includes tropical radiations coincident with the cooling and drying during the Oligocene–Miocene, the radiations during the establishment of land bridges and possible island hopping, between North and South America, and the increased aridity during the Andean uplift.

The genus *Bursera* as a unit is debated. Some authors consider it to be monophyletic, all members of the genus closely related among themselves, or paraphyletic—members of the genus have different origins. Some authors have included parts of *Bursera* (subgenus *Bullockia*) within its sister genus *Commiphora* (myrrh), while others like Becerra have consistently argued that it is a natural group. Becerra points to two *Bursera* speciation events: one during the uplift of the Sierra Madre Occidental (34–15 MYA), which gave rise to the northern species, and another related to the diversification of southern lineages during the creation of the Trans-Mexican Volcanic Belt (peaking about 13.5 MYA).[88]

Two remarkable ecological features of plant-animal interaction are found only in *Burseras*: the squirt defense mechanism, and the coadaptive relationships between *Bursera* and herbivorous chrysomelid beetles (*Blepharida*). All *Bursera* species have

88. Weeks et al. 2014; Becerra et al. 2012; Becerra 2003.

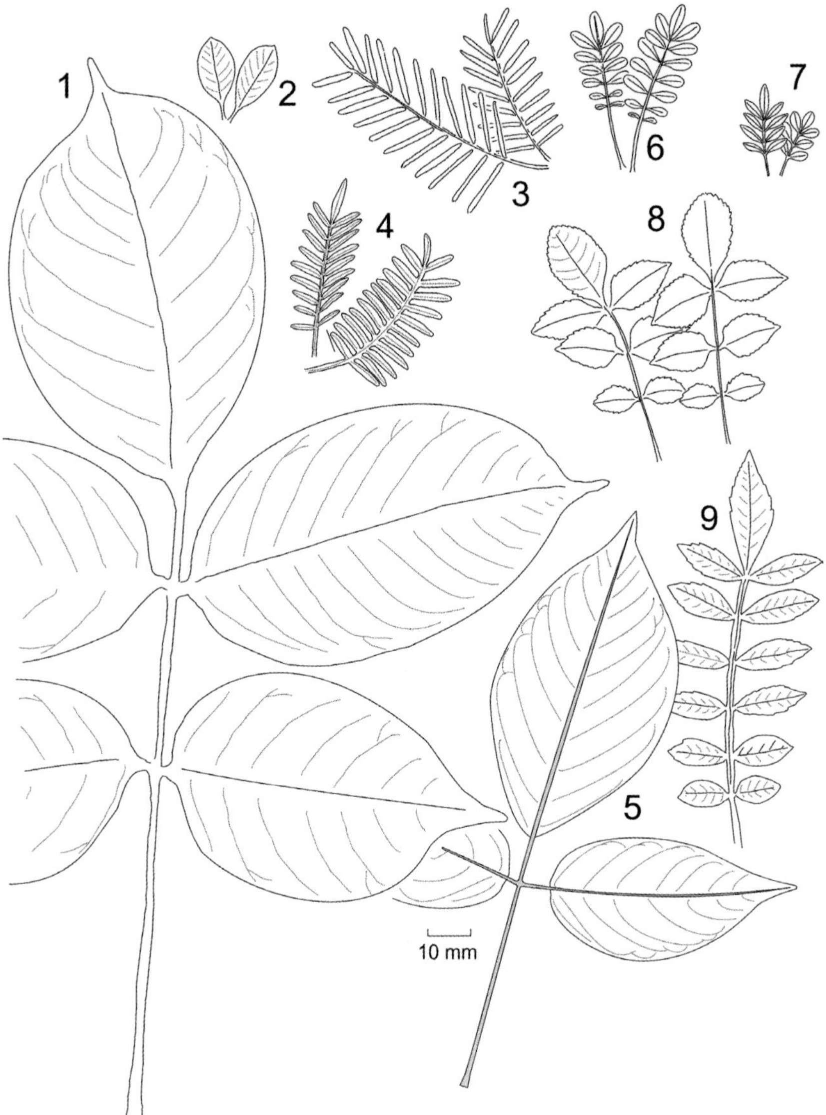

FIGURE 7.89 *Bursera* species with exfoliating bark in the Valleys of Tehuacán and Cuicatlán. Species with exfoliating reddish outer bark: (1) *B. simaruba*, (2) *B. schlechtendalii*, (3) *B. galeottiana*, (4) *B. morelensis*, and (5) *B. cinerea*. Species with exfoliating yellowish bark: (6) *B. aptera*, (7) *B. arida*, (8) *B. fagaroides*, and (9) *B. ariensis*. *B. simaruba* and *B. cinerea* can be taller than 20 meters. *B. morelensis* can be more than 10 meters tall. All other species are 2–8 meters. *Bursera arida* is a shrub 2–3 meters tall. All leaves drawn at about the same scale. Drawn from Rzedowski et al. (2004), photographs of live plants, and herbarium specimens.

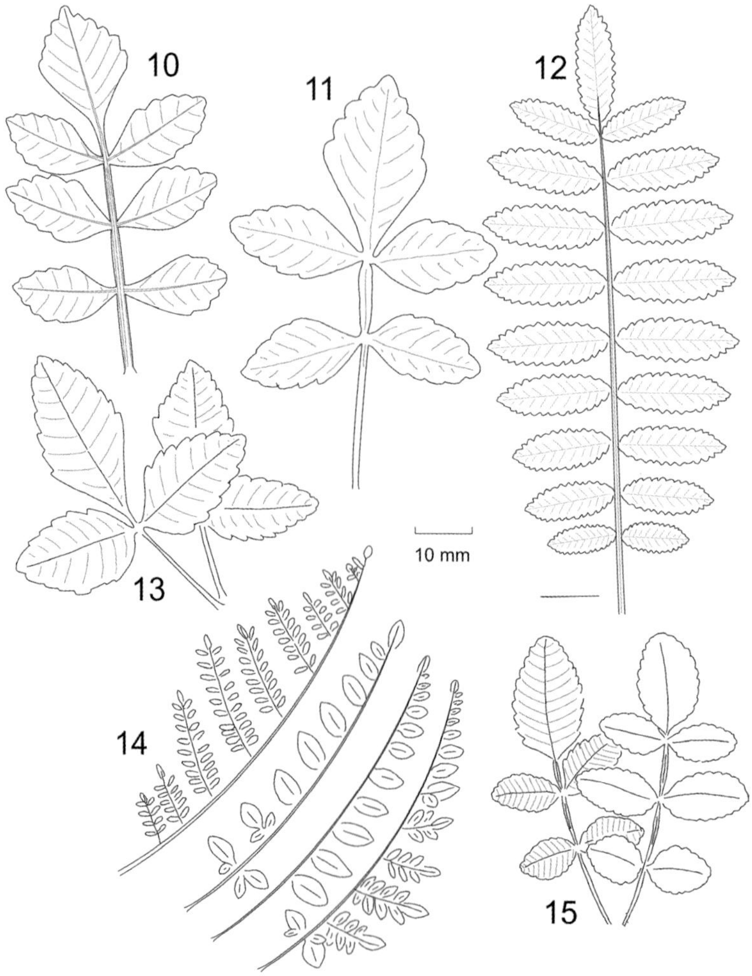

FIGURE 7.90 *Bursera* species with mostly nonexfoliating bark in the Valleys of Tehuacán and Cuicatlán. Species with nonexfoliating grayish to grayish red outer bark: (10) *B. glabrifolia*, (11) *B. linanoe*, (12) *B. aspleniifolia*, (13) *B. biflora* (slightly exfoliating grayish yellow bark), (14) *B. bipinnata*, and (15) *B. altijuga*. All leaves drawn at about the same scale. Drawn from Rzedowski et al. (2004), photographs of live plants, and herbarium specimens.

terpenes as an antiherbivore strategy. Some, however, have taken matters beyond the passive release of exudates and exhibit an ingenious defense: when herbivores puncture their tissues, particularly the petiole of the leaves, toxic terpenoid resin under pressure is forcibly ejected. Most insects cease their predation attempts immediately after piercing the leaf when they are doused with an instant bath of highly irritating turpentine. In the valleys, *B. schlechtendalii* is notable for its squirting defense, which is readily observable by squeezing the base of a fruit or simply cutting it with a fingernail. Blepharid jumping beetles, however, have evolved strategies to

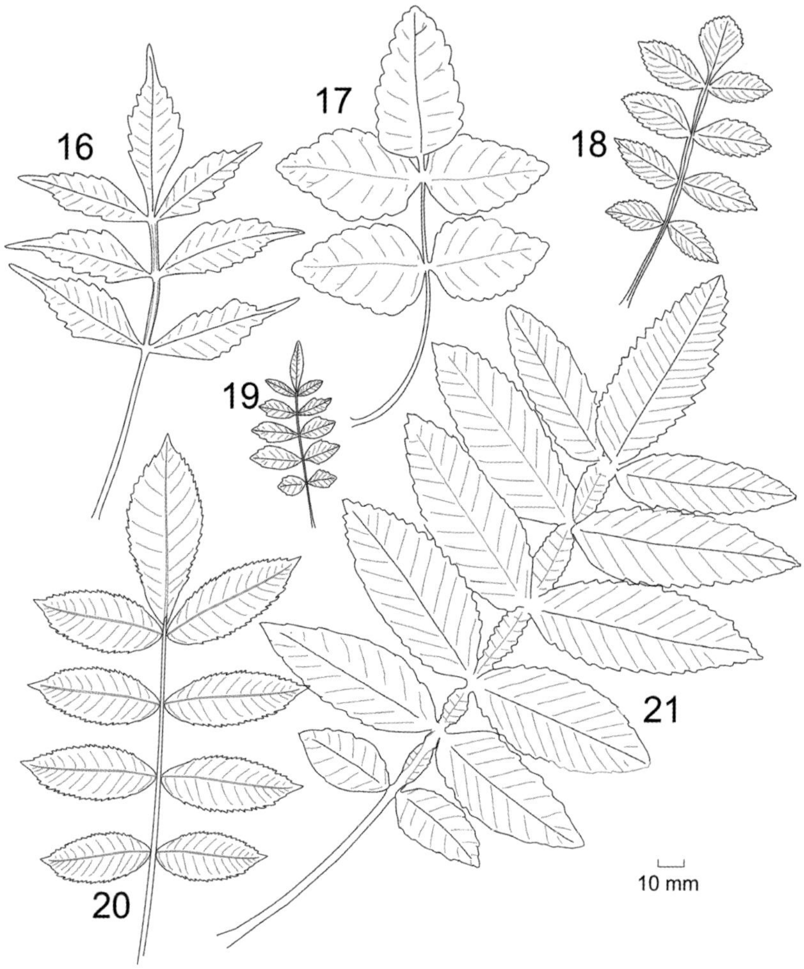

FIGURE 7.91 *Bursera* species with nonexfoliating grayish to grayish red outer bark in the Valleys of Tehuacán and Cuicatlán: (16) *B. graveolens*, (17) *B. pontiveteris*, (18) *B. submoniliformis*, (19) *B. mirandae*, (20) *B. esparzae*, and (21) *B. hintonii*. All leaves drawn at about the same scale. Drawn from Rzedowski et al. (2004), photographs of live plants, and herbarium specimens.

deal with the toxic resins of the plants and continue their predations undeterred. Because of the strong relationships and high specialization, *Bursera* and *Blepharida* are now considered a model system to study plant–herbivore coevolution.[89]

The proliferation of *Burseras*—and their odd characteristics—in TDF adds to the plant drama of the valleys. Those who have yet to experience a *cuajiotal* owe it to themselves to visit the subject of our final chapter, Cerro Petlanco.

89. Becerra et al. 2009; Becerra 1994, 2003.

CHAPTER 8

Cerro Petlanco

Heart of the Valleys

For our favorite location in the Cuicatlán and Tehuacán Valleys, we could have mentioned the cliffs overlooking Cuicatlán, the view of the lower valleys from Coxcatlán Cave, the interior of the forest of *tetechos* in the Helia Bravo Botanical Garden, the Cañada Obscura—dark gorge—of the Río Xiquila, or the view from atop Cerro Quiotepec. Instead, it is a humble, almost nondescript location near the southern end of the Valley of Tehuacán. Cerro Petlanco (*cerro* = hill in Spanish) is a most improbable candidate for international status among natural history sites. The hill, hardly more than 40 meters in height, rises gently from a semiarid plain, its topography barely discernible from more than a kilometer away and imperceptible from any but the highest resolution satellite images. Yet it occupies an important spot in local and regional legendry, has a deep archaeological history, and is a minor celebrity among plant enthusiasts.

We found that Cerro Petlanco embodies the multifaceted diversity of the valleys. Within only a few kilometers of the hill, one can recapitulate the valleys' complex natural history, their rich plant diversity, and their unusual geological features. Within a 25 kilometer radius, we find the valleys' most significant archaeological discoveries. To the west rise the wild slopes of the Mixteca Alta, mostly unpenetrated by roads; to the north loom the sierras of Zapotitlán. To the south, the Sierra Juárez and Sierra Monteflor merge with the Sierra Mixteca until they open into the central valleys of Oaxaca. To the east, the cliff-like limits of the Sierra Zongolica, the Sierra Mazateca, and the Sierra Juárez jut up from the valley bottoms. The hill itself and its immediate environs are rich in natural and human history.

Cerro Petlanco appears to straddle the Oaxaca-Puebla state line, though most of it lies within Oaxaca. (The exact location of the state line is and has been for many years subject to much local discussion.) It lies a few kilometers west of the

FIGURE 8.1 Sugarcane fields below forest of *tetechos*, Valley of Tehuacán.

Río Salado, on communal lands that belong to the San Gabriel Casablanca Indigenous Community.[1] The hill appears to have risen from the plains as a result of the flow of highly mineralized *manantiales*, artesian wells that for eons bubbled to the surface. Over the millennia, perhaps since the lake that covered the valleys was drained by the breaching of the Sierra Juárez by the Río Santo Domingo, the mineral-bearing waters that flowed to the surface evaporated, leaving behind a deposit that millimeter-by-millimeter grew into the hill that has become the Cerro Petlanco.

The waters have vanished, but as recently as 2006, according to residents of nearby villages, the mineral waters continued to bubble to the surface. Over the decades these gained considerable popularity for their reputed curative and restorative powers. Visitors from afar made the journey to soak in the waters, hoping to be relieved of bodily aches and pains and skin disorders, as well as to experience renewed youthful vigor. Residents note that pilgrims believed clothing drenched in the mineral waters and then dried would retain a curative virtue. Often the hillsides and top would be covered with an array of clothing spread out on the ground to dry, resembling a host of banners.

What caused the *manantiales*, the source of so much delight and relief to Oaxacans and Puebloans, to dry up as we now see them? Residents of adjacent villages propose two causes. One explanation, on which there is more or less universal agreement, is that blasting from a travertine (called *ónix* locally) quarry on a hill immediately to the west caused the collapse of the underground channels that guided the hydrothermal waters to the surface, thus sealing them off. In recent years, the community of San Gabriel Casablanca, which leases the quarry to the mining operator, has forbidden the use of explosives, but the damage is done.

The other theory for the disappearance of the springs, also widely believed, is that general overpumping of the regional aquifer has lowered the water table, removing the hydrostatic pressure that previously forced the mineral waters to the surface, as happens with the famed bubbling mineral springs near Tehuacán. Residents mention the thousands of acres of sugarcane fields to the north and east, most of which rely on vast quantities of pumped water for irrigation. Most of it is extracted upstream in the Río Salado drainage and delivered to the large plantations and to the ejidos by canals.

Both these theories are plausible. The hilltop of Cerro Petlanco exhibits dry declivities, with ghostly drainage holes in the bottom, where one day not too long

1. San Gabriel Casablanca has an unusual political structure—it is both an Indigenous community and an ejido, a rare phenomenon. Portions of the town lie in the state of Puebla, where ejidos are common. That portion is divided into both common lands and individually owned parcels, most of which are used for raising sugarcane. The lands in Oaxaca, where ejidos are few, are designated as *comunidades*, Indigenous lands that are communally owned, a symbol of the town's deep historical links to its Nahua ancestry.

FIGURE 8.2 David Yetman sits over a hole that until recently was filled with mineral water, Cerro Petlanco.

ago, visitors would sit, their feet and legs and perhaps their waist immersed in the warm waters that they believed would bestow healing powers. Archaeological studies demonstrate that the site was mined for its salts well before the arrival of Europeans, probably as far back as the twelfth century CE.[2] Pre-Columbians valued the salt and may also have viewed the waters as therapeutic. At the time of Conquest, Petlanco was probably included in the jurisdiction of Coxcatlán, which was a major exporter of salt, some of which would have come from Petlanco, some from nearby San Antonio Nanahuatípam.

Apart from the ongoing extractive work at the adjacent quarry, several thousand acres surrounding the hill are off limits to livestock grazing, woodcutting, hunting, and other forms of exploitation, because of their inclusion in the Cuicatlán-Tehuacán Biosphere Reserve. Enforcement of the conservation provisions lies with the leadership of the town of San Gabriel Casablanca. These individuals appear to be quite serious about protecting the hill and their portion of the reserve. Visitors are permitted inside the communal lands only with permission of the municipal president or the community's commissioner (*comisariado*), and then only accompanied by a designated guide from the village. In our case, Salvador Chávez, a cane farmer trained in the ecology of the thornscrub, proved a most knowledgeable guide.

2. MacNeish, Peterson, and Neely 1972, 491, 493; Sisson 1973, 84, 101.

Cerro Petlanco warrants protection, if only because it provides a glimpse of the vegetation of the Valley of Tehuacán as it must have been prior to widespread clearing and heavy pressure from livestock grazing. Botanists for several decades have recognized its distinctive characteristics and vegetation.[3] It is home to the most accessible population of *jarilla*, or *barrilito*, the rare *Fouquieria purpusii*. It is an odd shrub, which reaches little more than 3 meters in height, resembling nothing more than inverted parsnips. Scattered over the hillsides grow eighteen of these weird creations, which have escaped the depredations of plant collectors. These grow amid the far more common and widespread relative, *tlapacón* (*F. formosa*), strange in its own right. Townsfolk from San Gabriel Casablanca know the *jarilla* plants well and guard them diligently. These odd and rare plants appear to flourish on slopes of travertine origin.[4] A survey carried out for a mining enterprise on Cerro Prieto, a travertine hill some 5 kilometers southwest of Petlanco, notes an average of four plants per hectare on that hill, a density like that of Cerro Petlanco.[5]

The hill itself is anticlimactic, more a rise than a hill. The thin mineral soils result in vegetation that is sparse, spiny, and scrubby, contrasting notably with the rich variety and density of plants on the surrounding deep-soiled plains, including numerous tree species. The approach through the thornscrub and savanna, however, is exhilarating. The narrow dirt road, often impassable during the rainy months of June through September, passes through dense forests of *tetechos*—some of the finest cactus forests anywhere—as it undulates northwestward from San Gabriel Casablanca. The columnar cacti, including also *Pachycereus hollianus*, *P. weberi*, *Cephalocereus columna-trajani*, and *Pilosocereus chrysacanthus*, are interspersed with low trees, many of them leguminous, but mixed with other species as well, especially of the families Anacardiaceae and Burseraceae. Ground-hovering cacti, such as *Coryphantha* species, and several species of *Mammillaria* abound. Abundant also is the stinging nettle, *Cnidoscolus* species, with its lovely flowers and burning hairs. The removal of livestock by the Indigenous community since the early 2000s has resulted in a dense and exuberant growth of the understory unseen in the valleys for hundreds of years.

After perhaps 4 kilometers, the dirt track, maintained only by the passing of quarry trucks, crosses a gentle ridge and descends almost imperceptibly through increasingly savanna-like vegetation, dominated by a variety of leguminous trees,

3. See C. E. Smith 1965b, 129–30.

4. Gestión y Asesoría en Ingeniería Ambiental 2006, 76.

5. Members of the Casablanca community claim jurisdiction over the hill. Our first (failed) attempt to reach Cerro Petlanco was via San José Axuxco, Puebla, some 10 kilometers north. A resident purporting to be the owner of the hill asserted that he would have to guide us there (but was at that time unable to) and had strewn the road with nails to prevent unauthorized access to the hill.

FIGURE 8.3 Alberto Búrquez and *Fouquieria purpusii*.

FIGURE 8.4 *Mala mujer*, endemic stinging nettle (*Cnidoscolus tehuacanensis*), occupying a former warm spring site, Cerro Petlanco.

FIGURE 8.5 Valley of Tehuacán from Cerro Petlanco.

FIGURE 8.6 Road to Cerro Petlanco, San Gabriel Casablanca, Valley of Tehuacán

several species of *Burseras,* and three prominent members of the family Anacardiaceae: *cuachalalá* (*Juliana* [*Amphipterygium*] *adstringens*), whose bark is widely recognized for its curative properties; the *chupandilla* (*Cyrtocarpa procera*), with edible, plumlike fruits; and *Pseudosmodingium andrieuxii,* locally called *hinchahuevos* because of its notorious propensity for producing soreness, inflammation, and swelling of the testicles. It is a tree avoided by all knowledgeable men.[6]

The Burseras are more personable trees. The red-barked *B. morelensis* grows into an especially attractive 10-meter-tall tree with agreeably gnarled and thick branches. Other *Burseras* include *B. aptera, B. arida, B. biflora, B. fagaroides, B. graveolens, B. submoniliformis,* and *B. schlechtendalii,* the latter with its famous "squirt" defense.

And there are other notable tree species, many others, especially the tallest tree, the venerable *pochote, Ceiba aesculifolia.* Common also are the morning glory tree, *Ipomoea carnea,* and the ever-present palo verde (*Parkinsonia praecox*), along with a host of leguminous species. Columnar cacti, especially *Cephalocereus columna-trajani, Myrtillocactus geometrizans,* and *Neobuxbaumia tetetzo,* abound, poking their crowns above the canopy.

---

6. Eugene Hunn (2008, 69) mentions that the effects of *P. andrieuxii* are well known in the montane Zapotec community of San Juan Gbëë, in the montane forests of the Sierra Miahuatlán in Oaxaca.

FIGURE 8.7 *Pseudosmodingium andrieuxii*, Cerro Petlanco. It has unpleasant effects on humans.

FIGURE 8.8 Siesta time, Teotitlán de Flores Magón. Alberto Búrquez, whose photo is published here, claims nine dogs are visible. Skeptics count only eight, until they spot the upper dog with an apparent extra set of legs.

This is a most pleasant land, a vast plain rich with trees that regularly reach 10 meters in height, and during the rainy season, when they are fully leafed out, they provide rewarding shade (and menacing doses of chiggers that colonize the tall grasses, including the introduced buffelgrass).[7] The bark of the various *Bursera* species varies in color and texture, from the deep red of *B. morelensis* and *B. schlechtendalii* to green, gray, and brown, some species with bark that peels in sheets, some with silky smooth texture, characteristics that affirm their distinct evolutionary origins. While it is tempting to label this vegetation tropical deciduous forest, it is varied enough and has sufficient savanna-like properties that it more closely resembles the *caatinga* of the *sertão* of Brazil's Northeast, or the plains of the Sonoran Desert in the state of Sonora, than the dry tropical forests of Mexico's Pacific coast (see figure 7.84).

Cerro Petlanco, along with its environs, is a place to linger and savor the ambience—and the variety of plant shapes, sizes, textures, hues, and aromas—taking comfort in the apparent recognition by San Gabriel Casablanca residents that they hold the keys to a treasure, a relict of what was once the great wilderness of the Tehuacán and Cuicatlán Valleys.

---

7. Local experts discourage men from taking advantage of the shade of *Pseudosmodingium* or, for that matter, even approaching the tree.

APPENDIX

Species of Cactaceae Reported by BGIF for the Region of Tehuacan and Cuicatlán Valleys

+*Acanthocereus subinermis* Britton & Rose
Aporocactus martianus (Zucc.) Britton & Rose
+*Cephalocereus columna-trajani* (Karw. ex Pfeiff.) K. Schum.
Coryphantha calipensis Bravo ex S. Arias Montes, U. Guzmán Cruz & S. Gama López
Coryphantha pallida Britton & Rose
Coryphantha pseudoradians Bravo
Coryphantha retusa (Pfeiff.) Britton & Rose
Coryphantha retusa var. *melleospina* (Bravo) Bravo
Cylindropuntia imbricata (Haw.) F. M. Knuth
Cylindropuntia kleiniae (DC.) F. M. Knuth
Cylindropuntia leptocaulis (DC.) F. M. Knuth
Cylindropuntia rosea (DC.) Backeb.
Disocactus ackermannii (Lindl.) Barthlott
Disocactus speciosus (Cav.) Barthlott
Echinocactus platyacanthus Link & Otto
+*Escontria chiotilla* (F. A. C. Weber) Rose
Echinocereus pulchellus var. *acanthosetus* (Arias & Guzmán) W. Blum
Ferocactus flavovirens (Scheidw.) Britton & Rose
Ferocactus haematacanthus Bravo. ex Backeb. & F. M. Knuth
Ferocactus macrodiscus subsp. *macrodiscus* (Mart.) Britton & Rose
Ferocactus recurvus (Mill.) Borg
Ferocactus robustus (Pfeiff.) Britton & Rose
Hylocereus undatus (Haw.) Britton & Rose
+*Isolatocereus dumortieri* (Scheidw.) Backeb.
Mammillaria albilanata Backeb.

Mammillaria carnea Zucc. ex Pfeiff.
Mammillaria crucigera subsp. *crucigera* Mart.
Mammillaria discolor subsp. *ochoterenae* (Bravo) U. Guzmán
Mammillaria dixanthocentron Backeb. ex Mottram
Mammillaria duoformis R. T. Craig & E. Y. Dawson
Mammillaria flavicentra Backeb. ex Mottram
Mammillaria haageana Pfeiff.
Mammillaria hernandezii Glass & R. A. Foster
Mammillaria huitzilopochtli D. R. Hunt
Mammillaria karwinskiana Mart.
Mammillaria kraehenbuehlii (Krainz) Krainz
Mammillaria lanata (Britton & Rose) Orcutt
Mammillaria mystax Mart.
Mammillaria napina J. A. Purpus
Mammillaria oteroi Glass & R. A. Foster
Mammillaria pectinifera F. A. C. Weber
Mammillaria polyedra Mart.
Mammillaria schmollii (Bravo) Werderm. ex Backeb.
Mammillaria solisioides Backeb.
Mammillaria sphacelata Mart.
Mammillaria supertexta hort. ex Pfeiff.
Mammillaria tepexicensis J. Meyrán
Mammillaria uncinata Zucc. ex Pfeiff.
Mammillaria varieaculeata F. G. Buchenau
Mammillaria zephyranthoides Scheidw.
+*Myrtillocactus geometrizans* (Mart. ex Pfeiff.) Console
+*Myrtillocactus schenckii* (J. A. Purpus) Britton & Rose
+*Neobuxbaumia macrocephala* (F. A. C. Weber ex K. Schum.) E. Y. Dawson
+*Neobuxbaumia mezcalaensis* (Bravo) Backeb.
+*Neobuxbaumia tetetzo* (F. A. C. Weber ex K. Schum.) Backeb.
Nyctocereus serpentinus (Lag. & Rodr.) Britton & Rose
Opuntia auberi Pfeiff.
Opuntia cochenillifera (L.) Mill.
Opuntia decumbens Salm-Dyck
Opuntia depressa Rose
Opuntia ficus-indica (L.) Mill.
Opuntia huajuapensis Bravo
Opuntia hyptiacantha F. A. C. Weber
Opuntia lasiacantha Pfeiff.

Opuntia parviclada S. Arias & Gama
Opuntia pilifera F. A. C. Weber
Opuntia pubescens H. L.Wendl. ex Pfeiff.
Opuntia streptacantha Lem.
Opuntia tehuacana S. Arias & U. Guzmán
Opuntia tehuantepecana (Bravo) Bravo
Opuntia tomentosa Salm-Dyck
Opuntia velutina F. A. C. Weber
+*Pachycereus grandis* Rose
+*Pachycereus hollianus* (F. A. C. Weber) Buxb.
+*Pachycereus marginatus* (DC.) Britton & Rose
+*Pachycereus weberi* (J. M. Coult.) Backeb.
Peniocereus viperinus (F. A. C. Weber) Buxb.
Pereskiopsis rotundifolia (A. D. C.) Britton & Rose
+*Pilosocereus chrysacanthus* (F. A. C. Weber ex Schum.) Byles & G. D. Rowley
+*Polaskia chende* (Gosselin) A. C. Gibson & K. E. Horak
+*Polaskia chichipe* (Gosselin) Backeb.
+*Mitrocereus fulviceps* (F. A. C. Weber ex K. Schum.) Backeb. ex Bravo
Rhipsalis baccifera (Sol.) Stearn.
Selenicereus grandiflorus (L.) Britton & Rose
Selenicereus purpusii (Weing.) S. Arias & N. Korotkova
Stenocactus crispatus (DC.) A. Berger ex A. W. Hill
+*Stenocereus pruinosus* (Otto ex Pfeiff.) Buxb.
+*Stenocereus stellatus* (Pfeiff.) Riccob.
+*Stenocereus treleasei* (Vaupel) Backeb. (reputed)

+ = columnar species

Source: Salvador Arias Montes, *Actualización de la base de datos de Cactáceas del Valle de Tehuacán—Cuicatlán (Puebla, Oaxaca)*, version 1.7, occurrence dataset, Comisión Nacional para el Conocimiento y uso de la Biodiversidad C, Global Biodiversity Information Facility, May 30, 2020, https://doi.org/10.15468/d4tgq6.

REFERENCES

Acuña, A. M., L. Caso, M. M. Aliphat, and C. H. Vergara. 2011. "Edible Insects as Part of the Traditional Food System of the Popoloca Town of Los Reyes Metzontla, Mexico." *Journal of Ethnobiology* 31 (1): 150–69.

Acuña-Soto, Rodolfo, Leticia Calderón Romero, and James H. Maguire. 2000. "Large Epidemics of Hemorrhagic Fevers in Mexico 1545–1815." *American Journal of Tropical Medicine and Hygiene* 62 (6): 733–39.

Adams, Richard E. W. 1991. *Prehistoric Mesoamerica*. Norman: University of Oklahoma Press.

Aiuvalasit, Michael J., James A. Neely, and Mark D. Bateman. 2010. "New Radiometric Dating of Water Management Features at the Prehistoric Purrón Dam Complex, Tehuacán Valley, Puebla, México." *Journal of Archaeological Science* 37 (6): 1207–13.

Ajofrín, Fray Francisco de. 1959. *Diario del viaje que hicimos a México fray Francisco de Ajofrín y fray Fermín de Olite, capuchinos*. Madrid: Real Academia de la Historia.

Alaniz-Álvarez, Susana, and Ángel Nieto-Samaniego. 2007. *Geology of México: Celebrating the Centenary of the Geological Society of México*. Boulder, Colo.: Geological Society of America.

Arakaki, M., P.-A. Christin, Nyffeler, A. Lendel, U. Eggli, R. M. Ogburn, E. Spriggs, M. J. Moore, and E. J. Edwards. 2011. "Contemporaneous and Recent Radiations of the World's Major Succulent Plant Lineages." *Proceedings of the National Academy of Sciences* 108 (20): 8379–84. https://doi.org/10.1073/pnas.1100628108.

Arias Montes, Salvador. 2020. *Actualización de la base de datos de Cactáceas del Valle de Tehuacán—Cuicatlán (Puebla, Oaxaca)*. Version 1.7. Occurrence dataset. Comisión nacional para el conocimiento y uso de la biodiversidad C. Global Biodiversity Information Facility, May 30. https://doi.org/10.15468/d4tgq6.

Arias Montes, Salvador, Susana Gama López, and Leonardo U. Guzmán Cruz. 1997. *Flora del Valle de Tehuacán-Cuicatlán*, vol. 14, *Cactaceae A. L. Juss*. Mexico City: Universidad Nacional Autónoma de México.

Bakewell, Peter, and Jacqueline Holler. 2010. *A History of Latin America to 1825*. 3rd ed. London: Wiley-Blackwell.

Balkansky, Andrew. 1998. "Origin and Collapse of Complex Societies in Oaxaca (Mexico): Evaluating the Era from 1965 to the Present." *Journal of World Prehistory* 12 (4): 451–93.

Bandelier, A. 1884. "Report on an Archaeological Tour in Mexico in 1881." In *Papers of the Archaeological Institute of America, 1884*, 266–68. Boston, Mass.: Cupples, Upham.

Barba, Luis A., and José Luis Córdova Frunz. 1999. "Estudios energéticos de la producción de cal en tiempos teotihuacanos y sus implicaciones." *Latin American Antiquity* 10 (2): 168–79.

Becerra, Judith X. 1994. "Squirt-Gun Defense in *Bursera* and the Chrysomelid Counterploy." *Ecology* 75 (7): 1991–96.

Becerra, Judith X. 2003. "Evolution of Mexican *Bursera* (Burseraceae) Inferred from ITS, ETS, and 5S Nuclear Ribosomal DNA Sequences." *Molecular and Phylogenetic Evolution* 26 (2): 300–309.

Becerra, Judith X., K. Noge, S. Olivier, and D. L. Venable. 2012. "The Monophyly of *Bursera* and Its Impact for Divergence Times of Burseraceae." *Taxon* 61 (2): 333–43.

Becerra, Judith X., K. Noge, and D. L. Venable. 2009. "Macroevolutionary Chemical Escalation in an ancient Plant-Insect Arms Race." *Proc. Natl. Acad. Sci. USA* 106 (43): 18062–66.

Bharucha, Rustom. 2003. *Rajashtan, an Oral History: Conversations with Komal Kothari.* New Delhi: Penguin.

Blanco, A., G. Pérez, B. Rodríguez, N. Sugiyama, F. Torres, and R. Valadez. 2009. "El zoológico de Moctezuma ¿Mito o realidad? / Moctezuma's Zoo: Myth or Reality?" *AMMVEPE (Asociación Mexicana de Médicos Veterinarios Especialistas en Pequeñas Especies)* 20 (2): 28–39.

Black, Clanton C., and C. Barry Osmond. 2005. "Crassulacean Acid Metabolism Photosynthesis: 'Working the Night Shift.'" In *Discoveries in Photosynthesis*, edited by Govindjee, J. T. Beatty, H. Gest, and J. F. Allen, 881–93. Dordrecht: Springer.

Brunet, Jean. 1967. "Geologic Studies." In *The Prehistory of the Tehuacán Valley*, vol. 1, *Environment and Subsistence*, edited by D. S. Byers, 66–90. Austin: University of Texas Press.

Burgoa, Fray Francisco de. (1674) 1934. *Geográfica descripción de la parte Septentrional del polo ártico de la América y nueva iglesia de las indias occidentales y sitios astronómicos de esta provincia de predicadores de Antequera, Valle de Oaxaca.* Mexico City: Talleres Gráficos de la Nación.

Byers, Douglas. 1967. "Climate and Hydrology." In *The Prehistory of the Tehuacán Valley*, vol. 1, *Environment and Subsistence*, edited by D. S. Byers, 48–65. Austin: University of Texas Press.

Campos-Enríquez, J. O., J. Corbo-Camargo, J. Arzate-Flores, J. D. Keppie, C. Arango-Galván, M. Unsworth, and S. I. Belmonte-Jiménez. 2013. "The Buried Southern Continuation of the Oaxaca-Juarez Terrane Boundary and Oaxaca Fault, Southern Mexico: Magnetotelluric Constraints." *Journal of South American Earth Sciences* 43:62–73.

Cardel, Yuira, Victor Rico Gray, José García-Franco, and Leonard Thien. 1997. "Ecological Status of *Beaucarnea gracilis*, an Endemic Species of the Semiarid Tehuacán Valley, México." *Conservation Biology* 11:367–74.

Casas, A., A. Valiente-Banuet, J. L. Viveros, P. Dávila, R. Lira, and J. Caballero. 2001. "Plant Resources of the Tehuacán Valley, Mexico." *Economic Botany* 55:129–66.

Caso, Alfonso. 1963. "The Lords of Yanhuitlán." In *Ancient Oaxaca: Discoveries in Mexican Archeology and History*, edited by John Paddock, editor, 313–35. Palo Alto, Calif.: Stanford University Press.

Castellón Huerta, Blas Román. 2006. *Cuthá: El cerro de la máscara.* Arqueología y etnicidad en el sur de Puebla. Mexico City: INAH.

Castellón Huerta, Blas Román. 2009. *Un grano de sal: Aportaciones etnoarqueológicas al estudio histórico de una industria ancestral.* Mexico City: Universidad Nacional Autónoma de México.

Castillo Tejero, Noemí. 2000. "Trabajos arqueológicos recientes en Tehuacán, Puebla." In *Arqueología, historia y antropología: In memoriam José Luis Lorenzo Bautista*, edited by Jaime Litvak and Lorena Mirambell, 299–318. Mexico City: INAH, Colección Científica.

Castillo Tejero, Noemí. 2002. "Representaciones de astros en una vasija de Tehuacán, Puebla." In *Iconografía mexicana III: Los representantes de los astros*, edited by Beatriz Barba de Piña Chán, 73–81. Mexico City: INAH.

Castillo Tejero, Noemí. 2004. "Tepexi el Viejo, Puebla, un señorío popoloca del Posclásico en las fuentes y la arqueología." In *Homenaje a Jaime Litvak*, edited by Antonio Benevides, Linda Manzanilla, and Lorena Mirambell, 173–93. Mexico City: INAH, Colección Científica.

Castillo Tejero, Noemí. ca. 2008. "Proyecto sur del estado de Puebla área central Popoloca Tehuacán [*sic*]." *Anales de arqueología.* Mexico City: INAH.

Cepeda Cárdenas, Gerardo. 1997. "Los nonoalca y sus relaciones con los mixtecos en los siglos XI y XII." In *Simposium internacional Tehuacán y su entorno: balance y perspectivas*, edited by Eréndira de la Lama, 25–44. Mexico City: INAH.

Chadwick, Robert, and Richard MacNeish. 1967. "Codex Borgia and the Venta Salada Phase." In *The Prehistory of the Tehuacán Valley*, vol. 1, *Environment and Subsistence*, edited by D. S. Byers, 114–31. Austin: University of Texas Press.

Chang-Le, Ma, Robert Mosely, Chen Wen-Yun, and Zhou Zhe-Kun. 2007. "Plant Diversity and Priority Conservation Areas of Northwestern Yunnan, China." *Biodiversity and Conservation* 16:757–74.

Chemnick, Jeff. 2013. "The Dioons of Oaxaca." *Cactus and Succulent Journal* 85:19–27.

Cline, Howard F. 1963. "Colonial Mazatec Lienzos and Communities." In *Ancient Oaxaca*, edited by John Paddock, 270–97. Palo Alto, Calif.: Stanford University Press.

Códice Borgia. 1976. *Códice Borgia: Biblioteca Apostólica Vaticana (Cod. Borg. Messican 1).* Vollständige Faksimile-Ausgabe des Codex im Originalformat. Commentary by Karl Anton Nowotny. Graz, Austria: Akademische Druck- und Verlagsanstalt.

Códice Chimalpopoca. (1992) 2019. *Anales de Cuauhtitlan y Leyenda de los Soles.* Translated by Primo Feliciano Velázquez, with preface by Miguel León Portilla. Facsimile of 3rd ed. Mexico City: Universidad Nacional Autónoma de México. https://www.historicas.unam.mx/publicaciones/publicadigital/libros/000/codice_chimalpopoca.html.

Coe, Michael D. 1999. *The Maya: Ancient Peoples and Places.* London: Thames and Hudson.

Coe, Michael D., and Richard A. Diehl. 1980. *In the Land of the Olmec.* Austin: University of Texas Press.

Coe, Michael D., Dean Snow, and Elizabeth Benson. 1986. *Atlas of Ancient America.* New York: Facts on File.

Cook, Sherborne. 1947. "The Interrelation of Population, Food Supply, and Building in Pre-Conquest Central Mexico." *American Antiquity* 13 (1): 45–52.

Cortés, Hernán. 1963. *Cartas y Documentos: Porrúa #2*. Mexico City: Biblioteca Porrúa.

Crosby, Alfred. 2004. *Ecological Imperialism: The Biological Expansion of Europe, 900–1900*. Cambridge: Cambridge University Press.

Dávalos-Álvarez, Oscar Gabriel, Ángel Francisco Nieto-Samaniego, Susana A. Alaniz-Álvarez, Enrique Martínez-Hernández, and Elia Ramírez-Arriaga. 2007. "Estratigrafía cenozoica de la región Tehuacán y su relación con el sector norte de la falla de Oaxaca." *Revista Mexicana de Ciencias Geológicas* 24:197–215.

Davies, Nigel. 1973. *The Aztecs*. Norman: University of Oklahoma Press.

Dávila Aranda, Patricia, and Olga Herrera-MacBryde. 1997. "Tehuacán Cuicatlán Región." In *Centres of Plant Diversity: A Guide and Strategy for Their Conservation*, edited by S. D. Davis, V. H. Heywood, Olga Herrera-MacBryde, J. Villa-Lobos, and A. C. Hamilton, 139–43. Kew, Richmond, Surrey: The World Wide Fund for Nature (WWF) and IUCN: The World Conservation Union.

Dávila Aranda, Patricia, José Luis Villaseñor Ríos, Rosalinda Medina-Lémus, and Oswaldo Téllez Valdéz, eds. 1993–2012. *Flora del Valle de Tehuacán-Cuicatlán*. Mexico City: Universidad Nacional Autónoma de México, Instituto de Biología. http://www.ibiologia.unam.mx/barra/publicaciones/floras_tehuacan/florastehucan.htm.

Dávila Aranda, Patricia, María del Coro Arizmendi, Alfonso Valiente-Banuet, José Luis Villaseñor, Alejandro Casas, and Rafael Lira. 2002. "Biological Diversity in the Tehuacán-Cuicatlán Valley, Mexico." *Biodiversity and Conservation* 11:421–42.

Dean, W. R. J., and S. J. Milton. 2019. "The Dispersal and Spread of Invasive Alien *Myrtillocactus geometrizans* in the Southern Karoo, South Africa." *South African Journal of Botany* 121:210–15. https://doi.org/10.1016/j.sajb.2018.11.005.

De Benavente, Fray Toribio "Motolinía." 2014. *Historia de los Indios de la Nueva España*. Anejos de la Biblioteca Clásica de la Real Academia Española. Edición, estudio y notas de Mercedes Serna y Bernat Castany. Madrid: Real Academia Española Centro. https://www.fundacionaquae.org/wp-content/uploads/2017/07/Historia-de-los-Indios.pdf.

Del Castillo, Rafael, Jorge Pérez de la Rosa, Georgina Vargas Amado, and Raúl Rivera García. 2004. "Coníferas." In *Biodiversidad de Oaxaca*, edited by A. J. García-Mendoza, M. J. Ordóñez, and M. Briones-Salas, 141–58. Mexico City: Universidad Nacional Autónoma de México.

Delgado-Lemus, América, Ignacio Torres, José Blancas, and Alejandro Casas. 2014. "Vulnerability and Risk Management of *Agave* Species in the Tehuacán Valley, Mexico." *Journal of Ethnobiology and Ethnomedicine* 10 (53): 2.

De León, J. P. 2010. "Rethinking the Organization of Aztec Salt Production: A Domestic Perspective." *Archeological Papers of the American Anthropological Association* 19:45–57.

Del Paso y Troncoso, Francisco, ed. 1905–. *Papeles de la Nueva España*. 7 vols. Madrid: Est. Tipográfico Sucesores de Rivadeneyra.

De-Nova, J. A., Medina, R., Montero, J. C., Weeks, A., Rosell, J. A., Olson, M. E., Eguiarte, L. E. and S. Magallón. 2012. "Insights into the Historical Construction of Species-Rich Mesoamerican Seasonally Dry Tropical Forests: The Diversification of *Bursera* (Burseraceae, Sapindales)." *New Phytologist* 193 (1): 276–87. https://doi.org/10.1111/j.1469-8137.2011.03909.x.

Díaz del Castillo, Bernal. (1575) 1862–63. *Verdadera historia de los sucesos de la conquista de la Nueva-España*. 3 vols. Madrid. (Project Gutenberg, 2011, https://www.gutenberg.org/ebooks/author/35868.)

Diguet, León. 1906. "Contribution A L'Étude Géographique du Mexique Précolombien: Le Mixtécapan." *Journal de la Société des Américanistes*, n.s., 3 (1) : 15–43.

Doocy, Shannon, Amy Daniels, Shayna Dooling, and Yuri Gorokhovich. 2013. "The Human Impact of Volcanoes: A Historical Review of Events 1900–2009 and Systematic Literature Review." *PLoS Curr.*, April 16. DOI:10.1371/currents.dis.841859091a706efebf8a30f4ed7a19011.

Doolittle, William E. 1990. *Canal Irrigation in Prehistoric Mexico*. Austin: University of Texas Press.

Drennan, R. D., P. T. Fitzgibbons, and H. Dehn. 2000. "Imports and Export in Classic Mesoamerican Political Economy: The Tehuacán Valley and the Teotihuacán Obsidian Industry." In *The Ancient Civilizations of Mesoamerica*, edited by Michael Smith and Marilyn Masson, 174–90. Oxford: Blackwell.

Durán, Fray Diego. (1588) 1994. *The History of the Indies of New Spain*. Translated and annotated with an introduction by Doris Heyden. Norman: University of Oklahoma Press.

Edinger, Steven. 1996. *The Road from Mixtepec: A Southern Mexican Town and the United States Economy*. Fresno, Calif.: Asociación Cívica Benito Juárez.

Elías-Herrera, M., Ortega-Gutiérrez, F., Sánchez-Zavala, J.L., Macías-Romo, C., Ortega-Rivera, A. and A. Iriondo. 2007. "The Caltepec Fault Zone: Exposed Roots of a Long-Lived Tectonic Boundary between Two Continental Terranes of Southern Mexico." *Geological Society of America Special Papers* 422:317–42.

Evans, S. T. 2010. "The Garden of the Aztec Philosopher-King." In *Gardening—Philosophy for Everyone*, edited by D. O. O'Brien, 205–19. Oxford: Wiley-Blackwell.

Ezcurra, Exequiel. 1992. "Crecimiento y colapso en la Cuenca de Mexico." *Revista Ciencias* 25:13–27.

Fargher, Lane F., Richard E. Blanton, Verenice Y. Espinoza, John Millhauser, Nezahualcoyotl Xiuhtecutli, and Lisa Overholtzer. 2011. "Tlaxcallan: The Archaeology of an Ancient Republic in the New World." *Antiquity* 85 (327): 172–86.

Fiedel, Stuart. 1987. *Prehistory of the Americas*. Cambridge: Cambridge University Press.

Flannery, Kent. 1983. "Precolumbian Farming in the Valleys of Oaxaca, Nochixtlán, Tehuacán, and Cuicatlán: A Comparative Study." In *The Cloud People: Divergent Evolution of the Zapotec and Mixtec Civilizations*, edited by Kent Flannery and Joyce Marcus, 323–39. New York: Academic Press.

Furst, Jill Leslie. 1978. *Codex Vindobonensis Mexicanus I: A Commentary*. Albany: State University of New York, Institute for Mesoamerican Studies.

Galindo-Tovar, Maria Elena, Amaury M. Arzate-Fernández, Nisao Ogata-Aguilar, and Ivonne Landero-Torres. 2007. "The Avocado (*Persea americana, Lauraceae*) Crop in Mesoamerica: 10,000 Years of History." *Harvard Papers in Botany* 12:325–34.

Gámez Espinosa, Alejandra. 2003. *Los Popolocas de Tecamachalco-Quecholac: Historia, cultura y sociedad de un senorío prehispánico*. Puebla, Mexico: Benemérita Universidad Autónoma de Puebla.

Gámez Espinosa, Alejandra. 2008. "Los linajes popolocas en el mapa de Cuauhtinchan II." In *Mapa de Cuauhtinchan II: Entre la ciencia y lo sagrado*, edited by Tim Tucker and Arturo

Montero, 69–90. Mesoamerican Research Foundation. México. Mexico City: Universidad Nacional Autónoma de México.

García-Mendoza, Abisaí J. 2004. "Agaváceas." In *Biodiversidad de Oaxaca*, edited by Abisaí García-Mendoza, María de Jesús Ordóñez, and Miguel Briones-Salas, 159–70. Mexico City: Universidad Nacional Autónoma de México.

García-Mendoza, Abisaí J., and Jorge A. Meave. 2011. *Diversidad florística de Oaxaca: De musgos a angiospermas (colecciones y lista de especies)*. Mexico City: Universidad Nacional Autónoma de México y Comisión Nacional para el Conocimiento y Uso de la Biodiversidad.

Gentry, Howard Scott. 1982. *Agaves of Continental North America*. Tucson: University of Arizona Press.

Gestión y Asesoría en Ingeniería Ambiental. 2006. *Manifestación de impacto ambiental en su modalidad particular para la explotación de mármol travertino en una mina a cielo abierto en la localidad conocida como Cerro Prieto ubicado en terrenos ejidales de San Antonio Nanahuatípam, municipio del mismo nombre, distrito de Teotitlán de Flores Magón, Oaxaca*. Gestión y Asesoría en Ingeniería Ambiental, September. http://sinat.semarnat.gob.mx/dgiraDocs/documentos/oax/estudios/2006/20OA2006M0004.pdf.

Golitko, M., and G. M. Feinman. 2015. "Procurement and Distribution of Pre-Hispanic Mesoamerican Obsidian 900 BC–AD 1520: A Social Network Analysis." *Journal of Archaeological Method and Theory* 22:206–47.

Gorenstein, Shirley. 1973. "Tepexi el Viejo: A Postclassic Fortified Site in the Mixteca-Puebla Region of Mexico." *Transactions of the American Philosophical Society* 63:1–75.

Guerrero-Arenas, Rosalía, Eduardo Jiménez Hidalgo, and Héctor Santiago Romero. 2010. "La transformación de los ecosistemas de la Mixteca Alta oaxaqueña desde el Pleistoceno Tardío hasta el Holoceno." *Ciencia y Mar* 14 (40): 61–68.

Hämäläinen, Pekka. 2022. *Indigenous Continent*. New York: Liveright.

Hapka, Roman, and Fabienne Rouvinez. 1997. "Las Ruin Cave, Cerro Rabón, Oaxaca, Mexico: A Mazatec Postclassic Funerary and Ritual Site." *Journal of Cave and Karst Studies* 59 (1): 22–25.

Hassig, Ross. 1988. *Aztec Warfare: Imperial Expansion and Political Control*. Norman: University of Oklahoma Press.

Hassig, Ross. 1992. *War and Society in Ancient Mesoamerica*. Berkeley: University of California Press.

Hernández, Sergio. 2014. "El ferrocarril que soñó Matías Romero." *Mexico Desconocido*, December.

Hipp, Andrew L., Paul S. Manos, Marlene Hahn, Michael Avishai, Catherine Bodénès, Jeannine Cavender-Bares, Andrew A. Crowl, Min Deng, Thomas Denk, Sorel Fitz-Gibbon, Oliver Gailing, M. Socorro González-Elizondo, Antonio González-Rodríguez, Guido W. Grimm, Xiao-Long Jiang, Antoine Kremer, Isabelle Lesur, John D. McVay, Christophe Plomion, Hernando Rodríguez-Correa, Ernst-Detlef Schulze, Marco C. Simeone, Victoria L. Sork, and Susana Valencia-Avalos. 2020. "Genomic Landscape of the Global Oak Phylogeny." *New Phytologist* 226 (4): 1198–1212. https://doi.org/10.1111/nph.16162.

Hopkins, Nicholas A. 1984. "Otomanguean Linguistic Prehistory." In *Essays in Otomanguean History*, edited by J. Kathryn Josserand, Marcus Winter, and Nicholas Hopkins, 25–64.

Vanderbilt University Publications in Anthropology No. 31. Nashville: Vanderbilt University, Department of Sociology and Anthropology.

Hose, Louise. 1995. "World's Deepest Karst Hydraulic System Documented in the Sierra Juárez, Oaxaca." *Revista Mexicana de Ciencias Geológicas* 12 (1): 108–11.

Hunn, Eugene. 2008. *A Zapotec Natural History*. Tucson: University of Arizona Press.

Hunt, Eva. 1972. "Irrigation and the Socio-Political Organization of Cuicatec Cacicazgos." In *The Prehistory of the Tehuacán Valley*, vol. 4, edited by Frederick Johnson, 162–259. Austin: University of Texas Press.

Hurd, Willis E. 1929. "Northers of the Gulf of Tehuantepec." *Monthly Weather Review* (American Meteorological Society) 57:192–94.

Ibarra Contreras, Clarita A. 2013. "Yanhuitlán: Un pueblo mixteco de tequio y conservación." *Revista Vinculando*, September 10. http://vinculando.org/articulos/sociedad_mexico/yanhuitlan-un-pueblo-mixteco-de-tequio-y-conservacion.html.

Instituto Nacional de Estadística y Geografía. 2010. *Principales resultados por localidad*. Mexico City: INEGI.

Jäcklein, Klaus J. 1979. "Apuntes sobre la historia prehispánica de los Popolocas de Puebla." In *Mesoamerica: Homenaje al Doctor Paul Kirchhoff*, edited by Barbro Dalhgren, 194–211. Mexico City: INAH.

Jansen, Maartin. 1998. *The Shadow of Monte Albán: Politics and Historiography in Postclassic Oaxaca, Mexico*. Leiden: Research School CNWS, School of Asian, African and Amerindian Studies.

Jiménez-Barron, Ofelia, Ricardo García-Sandoval, Susana Magallón, Abisaí J. García-Mendoza, Jorge Nieto-Sotelo, Erika Aguirre-Planter, and Luis E. Eguiarte. 2020. "Phylogeny, Diversification Rate, and Divergence Time of *Agave* Sensu Lato (Asparagaceae), a Group of Recent Origin in the Process of Diversification." *Frontiers in Plant Science* 11:1651. https://doi.org/10.3389/fpls.2020.536135.

Jiménez Moreno, Wigberto. 1966. "Mesoamerica before the Toltecs." In *Ancient Oaxaca: Discoveries in Mexican Archeology and History*, edited by John Paddock, 1–82. Palo Alto, Calif.: Stanford University Press.

Johnson, Frederick, ed. 1972. *Prehistory of the Tehuacan Valley*. Vol. 4, *Chronology and Irrigation*. Austin: University of Texas Press.

Johnson, Nicolás. 1997. "The Route from the Mixteca Alta into Southern Puebla on the *Lienzo of Tlapiltepec*." In *Códices y documentos sobre México: Segundo Simposio*, vol. 1, edited by Constanza Vega Sosa and S. Rueda Smithers, 233–68. Mexico City: INAH.

Keppie, J. D. 2004. "Terranes of Mexico Revisited: A 1.3 Billion Year Odyssey." *International Geology Review* 46:765–94.

Kirby, Michael. 1972. *The Physical Environment of the Nochixtlán Valley, Oaxaca*. Vanderbilt University Publications in Anthropology 2. Nashville, Tenn.: Vanderbilt University.

Kowalewski, Stephen, Andrew Balkansky, Laura Stiver Walsh, Thomas Pluckhahn, John Chamblee, Verónica Pérez Rodríguez, Verenice Heredia Espinoza, and Charlotte Smith. 2009. *Origins of the Ñuu: Archaeology in the Mixteca Alta, Mexico*. Boulder: University of Colorado Press.

Kraft, Kraig H., Cecil H. Brown, Gary P. Nabhan, Eike Luedeling, José de Jesús Luna Ruiz, Geo Coppens d'Eeckenbrugge, Robert J. Hijmans, and Paul Gepts. 2014. "Multiple Lines of Evidence for the Origin of Domesticated Chili Pepper, *Capsicum annuum*, in Mexico." *Proceedings of the National Academy of Sciences* 111 (17): 6165–70. https://doi.org/10.1073/pnas.1308933111.

Krasilnikov, Pavel, Ma. del Carmen Gutiérrez-Castorena, Robert J. Ahrens, Carlos Omar Cruz-Gaistardo, Sergey Sedov, and Elizabeth Solleiro-Rebolledo. 2013. *Soils of Mexico.* Dordrecht: Springer Science and Business Media. Ebook.

Le Bot, Jean-Michel. 2012. "Contribution à l'histoire d'un lieu commun: l'attribution à Chateaubriand de la phrase 'les forêts précèdent les peuples, les déserts les suivent.'" *Sociologos* 7. http://journals.openedition.org/socio-logos/2634.

Leigh, David S., Stephen A. Kowalewski, and Genevieve Holdridge. 2013. "3400 Years of Agricultural Engineering in Mesoamerica: Lama-bordos of the Mixteca Alta, Oaxaca, Mexico." *Journal of Archaeological Science* 40 (11): 4107–11.

León, Nicolás. 1905. *Los Popolocas.* Mexico City: Fundación Amparo.

Levine, Marc N. 2011. "Negotiating Political Economy at Late Post Classic Tututepec (Yucu Dzaa), Oaxaca, Mexico." *American Anthropologist* 113:22–39.

Lewis, M. Paul, Gary F. Simons, and Charles D. Fanning, eds. 2013. *Ethnologue: Languages of the World.* 17th ed. Dallas, Tex.: Summer Institute of Linguistics International. http://www.ethnologue.com.

Lister, Robert, and Florence Lister. 1981. *Chaco Canyon: Archaeology and Archaeologists.* Albuquerque: University of New Mexico Press.

Long Towell, J., and A. Attolini Lecón. 2010. *Caminos y mercados de México.* Serie Historia General 23. Mexico City: Universidad Nacional Autónoma de México, Instituto de Investigaciones Históricas/INAH.

Long, Austin, and Gayle Fritz. 2001. "Validity of AMS Dates on Maize from the Tehuacán Valley: A Comment on MacNeish and Eubanks." *Latin American Antiquity* 12 (1): 87–90.

López, L. 1993. *Las ofrendas del Templo Mayor de Tenochtitlán.* Mexico City: SEP-INAH.

López-Galindo, F., D. Muñoz-Iniestra, M. Hernández-Moreno, A. Soler-Aburto, M. C. Castillo-López, and I. Hernández-Arzate. 2003. "Análisis integral de la toposecuencia y su influencia en la distribución de la vegetación la degradación del suelo en la subcuenca de Zapotitlán Salinas, Puebla." *Boletín de la Sociedad Geológica Mexicana* 56:19–41.

López de Gómara, Francisco. (1552) 1979. *Historia general de las Indias y Vida de Hernán Cortés.* Caracas: Fundación Biblioteca Ayacucho. http://biblioteca.clacso.edu.ar/clacso/se/20190905104522/Historia_General_de_las_Indias_y_vida_de_Hernan_Cortes_Francisco_Lopez_de_Gomara.pdf.

Macías Goytia, Angelina. 2000. "Implicaciones de fechamientos en concha." In *Arqueología, historia y antropología: In memoriam José Luis Lorenzo Bautista*, edited by Jaime Litvak and Lorena Mirambell, 373–88. Mexico City: INAH, Colección Científica.

MacNeish, Richard. 1964. "Ancient American Civilization." *Science* 143:531–37.

MacNeish, Richard. 1967. "An Interdisciplinary Approach to an Archaeological Problem" In *The Prehistory of the Tehuacán Valley*, vol. 1, *Environment and Subsistence*, edited by D. S. Byers, 14–24. Austin: University of Texas Press.

MacNeish, Richard. 1967. "A Summary of Subsistence." In *The Prehistory of the Tehuacán Valley*, vol. 1, *Environment and Subsistence*, edited by D. S. Byers, 290–309. Austin: University of Texas Press.

MacNeish, Richard, Frederick Peterson, and James Neely. 1972. "The Archaeological Reconnaissance." In *The Prehistory of the Tehuacán Valley*, vol. 5, *Excavations and Reconnaissance*, edited by R. S. MacNeish, 341–496. Austin: University of Texas Press.

Magaña, Víctor O., Jorge L. Vázquez, José L. Pérez, and Joel B. Pérez. 2003. "Impact of El Niño on Precipitation in Mexico." *Geofísica Internacional* 42 (3): 313–30.

Manin, Aurélie, and Christine Lefèvre. 2016. "The Use of Animals in Northern Mesoamerica, Between the Classic and the Conquest (200–1521 AD): An Attempt at Regional Synthesis on Central Mexico." *Anthropozoologica* 51 (2): 127–47. http://doi.org/10.5252/az2016n2a5.

Manzanilla, Linda. 2001. "Teotihuacán." In *The Oxford Encyclopedia of Mesoamerican Cultures*, vol. 3, edited by David Carrasco, 201–8. Oxford: Oxford University Press.

Manzanilla, Linda, Xim Bokhimi, Dolores Tenorio, Melania Jiménez-Reyes, Edgar Rosales, Cira Martínez, and Marcus Winter. 2017. "Procedencia de la mica de Teotihuacan: Control de los recursos suntuarios foráneos por las élites gobernantes." *Anales de Antropología* 51:23–38.

Marcus, Joyce. 2000. "On the Nature of the Mesoamerican City." In *The Ancient Civilizations of Mesoamerica*, edited by Michael E. Smith and Marilyn Masson, 49–82. Oxford: Oxford, Blackwell.

Marcus, Joyce, and Kent V. Flannery. 1996. *Zapotec Civilization: How Urban Society Evolved in Mexico's Oaxaca Valley*. London: Thames and Hudson.

Marcus, Joyce, and Kent V. Flannery. 2000. "Cultural Evolution in Oaxaca: The Origins of the Zapotec and Mixtec Civilizations." In *The Cambridge History of the Native Peoples of the Americas*, vol. 2, *Mesoamerica, Part I*, edited by Richard Adams and Murdo MacLeod, 358–406. Cambridge: Cambridge University Press.

Martin, Paul, David Yetman, Mark Fishbein, Phil Jenkins, Thomas Van Devender, and Rebecca Wilson. 2000. *Gentry's Rio Mayo Plants: The Tropical Deciduous Forest and Environs of Northwest Mexico*. Tucson: University of Arizona Press.

Martin, Paul S., and Richard Klein. 1984. *Quaternary Extinctions: A Prehistoric Revolution*. Tucson: University of Arizona Press.

Martínez Donjuán, Guadalupe. 2001. "Teopantecuanitlán." In *The Oxford Encyclopedia of Mesoamerican Cultures*, vol. 3, edited by David Carrasco, 200–201. Oxford: Oxford University Press.

Mastache, Alba G., and Robert H. Cobean. 2001. "Tula." In *The Oxford Encyclopedia of Mesoamerican Cultures*, vol. 3, edited by David Carrasco, 268–74. Oxford: Oxford University Press.

McCafferty, Geoffery. 2001. "Cholula." In *The Oxford Encyclopedia of Mesoamerican Cultures*, vol. 1, edited by David Carrasco, 202–6. Oxford: Oxford University Press.

McKillop, Heather. 2017. "Early Maya Navigation and Maritime Connections in Mesoamerica." In *The Sea in History—The Medieval World*, edited by C. Buchet and M. Balard, 701–15. Woodbridge: Boydell and Brewer. https://doi.org/10.1017/9781782049104.062.

Medina-Sánchez, Javier, Sue J. McLaren, José Ortega-Ramírez, and Alfonso Valiente-Banuet. 2020. "Tectono-Stratigraphic Basin Evolution in the Tehuacán-Mixteca Highlands, Southwestern México." *Heliyon* 6 (3): e03584.

Melville, Eleanor. 1994. *A Plague of Sheep: Environmental Consequences of the Conquest of Mexico*. Cambridge: Cambridge University Press.

Mendoza García, Edgar. 2002. "El ganado comunal en la Mixteca Alta: De la época colonial al siglo XX. El caso de Tepelmeme." *Historia Mexicana* 51 (4): 749–85. http://www.redalyc.org/articulo.oa?id=60051402.

Michalzik, Dieter, Rudolf Fischer, Delfino Hernández, and Deniz Oezen. 2001. "Age and Origin of the 'Mexican Onyx' at San Antonio Texcala (Puebla, Mexico)." *Geologische Beiträge Hannover* 2:79–89.

Michan Aguirre, L., and J. Llorente Bousquets. 2003. *La taxonomía en México durante el siglo XX.* Publicaciones Especiales del Museo de Zoología 12. Mexico City: Facultad de Ciencias, Universidad Nacional Autónoma de México.

Millon, Rene. 1976. "Social Relations in Ancient Teotihuacán." In *The Valley of Mexico: Studies in Prehispanic Ecology and Society,* edited by Eric Wolf, 205–48. Albuquerque: University of New Mexico Press.

Miranda, Faustino. 1942. "Estudios sobre la vegetación de México III: Notas generales sobre la vegetación del suroeste del estado de Puebla, especialmente en la zona de Izúcar de Matamoros." *Anales del Instituto de Biología Universidad Nacional Autónoma de México* 13 (2): 417–59.

Miranda, Faustino. 1948. "Datos sobre la vegetación en la cuenca alta del Papaloapan." *Anales del Instituto de Biología Universidad Nacional Autónoma de México* 19:333–64.

Miranda, Faustino, and Efraím Hernández-Xolocotzi. 1963. "Los tipos de vegetación de México y su clasificación." *Boletín de la Sociedad Botánica de Mexico* 28:29–179.

Morales Folguera, José Miguel. 2004. "Jardines prehispánicos de México en las crónicas de Indias." *AEA (Archivo Español de Arte)* 77 (308): 351–73.

Morante López, Rubén. 2010. "Las antiguas rutas comerciales: Un camino por las sierras nahuas de Puebla y Veracruz." In *Caminos y Mercados de México,* edited by Janet Long Towell and Amalia Attolini Lecón, 107–27. Mexico City: Universidad Nacional Autónoma de México.

Morrill, Penny. 2008. "Roads and Bridges in the Ancient Americas." In *Encyclopedia of Society and Culture in the Ancient World,* edited by Peter Bogucki, 891–93. New York: Facts on File.

Narváez Carvajal, G. 1991. "Las condiciones naturales para la producción agrícola en las mixtecas oaxaqueñas." *Revista de Geografía Agrícola* 15/16:33–40.

Navarrete Linares, Federico. 2011. *Los orígenes de los orígenes de los pueblos indígenas del Valle de México: Los altéptl y sus historias.* Mexico City: Universidad Nacional Autónoma de México, Instituto de Investigaciones Históricas. http://www.historicas.unam.mx/publicaciones/publicadigital/libros/origenes/origenespueblos.html.

Neall, V. E. 2009. "Volcanic Soils." *Encyclopedia of Land Use, Land Cover and Soil Sciences,* vol. 7, *Soils and Soil Sciences (Part 2),* edited by Willy H. Verheye, 23–45. Oxford: Encyclopaedia of Life Support Systems (EOLSS).

Negra, Debra. 1989. "Public Proclamation in the Art of Cacaxtla and Xochicalco." In *Mesoamerica After the Decline of Teotihuacan A.D. 700–900,* edited by Richard Diehl and Janet Berlo, 83–104. Washington D.C.: Dumbarton Oaks Research Library and Collection.

Nichols, Johanna. 1992. *Linguistic Diversity in Time and Space.* Chicago: University of Chicago Press.

Nicholson, Rob. 2002. "The Splendid Haul of Cyrus Guernsey Pringle." *Arnoldia* 61 (2): 2–13.

Nixon, Kevin C. 2006. "Global and Neotropical Distribution and Diversity of Oak (Genus *Quercus*) and Oak Forests." In *Ecology and Conservation of Neotropical Montane Oak Forests*, edited by M. Kappelle, 3–13. Ecological Studies 185. Berlin: Springer-Verlag.

Noble, David G. 2004. *In Search of Chaco*. Santa Fe, N.Mex.: School of American Research Press.

Nuttall, Zelia. 1974. *Códice Nuttall: Reproducción del facsímile editado por El Museo Peabody de la Universidad de Harvard*. Mexico City: La Estampa Mexicana.

Padden, R. C. 1967. *The Hummingbird and the Hawk*. New York: Harper and Row.

Paddock, John. 1966. *Ancient Oaxaca*. Palo Alto, Calif.: Stanford University Press.

Paddock, John. 1997. "Algunos problemas pendientes en el Valle de Tehuacán." In *Simposium internacional Tehuacán y su entorno: Balance y perspectivas*, edited by Eréndira de la Lama, 103–8. Mexico City: INAH.

Palerm, Angel, Juan Maestre Alfonso, and Alba González Jácome. 2007. *Agua y agricultura: Ángel Palerm, la discusión con Karl Wittfogel sobre el modo asiático de producción y la construcción de un modelo para el estudio de Mesoamérica*. Mexico City: Universidad Iberoamericana.

Paredes Colín, Joaquín. (1910) 1953. *Apuntes históricos de Tehuacán*. Tehuacán, Puebla, Mexico: Diseños e Impresos Tehuacán.

Parsons, H. R. 1989. "Una etnografía arqueológica de la producción tradicional de sal en Nexquipayac, Estado de México." *Arqueología* 2:69–80.

Pastrana, Alejandro, and Patricia Fournier. 2001. "Cuicuilco." In *The Oxford Encyclopedia of Mesoamerican Cultures*, vol. 1, edited by David Carrasco, 290–92. Oxford: Oxford University Press.

Pérez Malváez, Carlos, David N. Espinosa-Organista, and Antonio A. Bueno-Hernández. 2007. "Las ideas biogeográficas de Faustino Miranda." In *Faustino Miranda: Una vida dedicada a la botánica*, edited by Francisco Javier Docil Mancilla, 232–42. Morelia: Universidad Michoacana San Nicolás de Hidalgo.

Pérez Rodríguez, Verónica. 2013. "Recent Advances in Mixtec Archaeology." *Journal of Archaeological Research* 21:75–121.

Pérez Rodríguez, Verónica, and Kirk C. Anderson. 2013. "Terracing in the Mixteca Alta, Mexico: Cycles of Resilience of an Ancient Land-Use Strategy." *Human Ecology* 41:335–49.

Piperno, Dolores, and Kent V. Flannery. 2001. "The Earliest Archaeological Maize (*Zea maize* L.) from Highland Mexico: New Accelerator Mass Spectrometry Dates and Their Implications." *Proceedings of the National Academy of Sciences* 98 (4): 2101–3.

Piperno, Dolores, Anthony Ranere, Irene Holst, José Iriarte, and Ruth Dickau. 2009. "Starch Grain and Phytolith Evidence for Early Ninth Millennium B.P. Maize from the Central Balsas River Valley, Mexico." *Proceedings of the National Academy of Sciences* 106 (13): 5109–24.

Plunket, Patricia, and Gabriela Uruñuela. 2005. "Recent Research in Puebla History." *Journal of Archaeological Research* 13 (2): 89–127.

Pohl, John. 1999. *Exploring America*. Oxford: Oxford University Press.

Ramírez, Alfonso Francisco. 1958. *Por los Caminos de Oaxaca*. Mexico City: Talleres Gráficos de la Nación.

Ramírez-Arriaga, E., M. B. Prámparo, A. F. Nieto-Samaniego, E. Martínez-Hernández, A. Valiente-Banuet, C. Macías-Romo, and O. G. Dávalos-Álvarez. 2014. "Palynological

Evidence for Middle Miocene Vegetation in the Tehuacán Formation of Puebla, Mexico." *Palynology* 38 (1): 1–27. http://dx.doi.org/10.1080/01916122.2013.802750.

Ramos-Elorduy, Julieta, and José Manuel Pino-Moreno. 1989. *Los Insectos comestibles en el México Antiguo: Estudio etnoentomológico.* Mexico City: AGT Editores.

Ramos-Elorduy Julieta, and José Manuel Pino-Moreno. 2005. *Base de datos de los insectos comestibles de México.* Mexico City: Edición Especial del Instituto de Biología, Universidad Nacional Autónoma de México.

Rattray, Evelyn Childs. 1990. "New Findings on the Origin of Thin Orange Ceramics." *Ancient Mesoamerica* 1:181–95.

Redmond, Elsa. 1983. *A Fuego y Sangre: Early Zapotec Imperialism in the Cuicatlán Cañada, Oaxaca.* Memoirs of the Museum of Anthropology, 16. Ann Arbor: University of Michigan, Museum of Anthropology.

Reff, Daniel. 1991. *Disease, Depopulation, and Cultural Change in Northwestern New Spain, 1518–1764.* Salt Lake City: University of Utah Press.

Reina Aoyama, Leticia. 1998. "Las rutas de Oaxaca." *Dimension Antropológica* 12:49–76.

Restall, Matthews. 2018. *When Montezuma Met Cortés: The True Story of the Meeting That Changed History.* New York: HarperCollins.

Reyes Santiago, Jerónimo, Christian Brachet, Joel Pérez Crisanto, and Araceli Gutiérrez de la Rosa. 2004. *Cactáceas y otras plantas nativas de La Cañada, Cuicatlán, Oaxaca.* Mexico City: Sociedad Mexicana de Cactología.

Reynoso, Luisa. 1997. "La cerámica de Los Reyes Metzontla, una clave hacia el pasado." In *Simposio internacional Tehuacán y su entorno: balance y perspectivas,* edited by Eréndira de la Lama, 115–28. Mexico City: INAH.

Rincón Mautner, Carlos. 2015. "Pluri-Ethnic Coixtlahuaca's Longue Durée." In *Bridging the Gaps: Integrating Archaeology and History in Oaxaca, Mexico,* edited by Danny Zborover and Peter Kroefges, 157–207. Boulder: University Press of Colorado.

Rivas Guevara, María, Benito Rodríguez Haros, and Jacinta Palerm Viqueira. 2008. "El sistema de jollas una técnica de riego no convencional en la Mixteca [*sic*]." *Boletín del Archivo Histórico del Agua,* no. 40, 6–16. https://dialnet.unirioja.es/servlet/articulo?codigo=3168668.

Rodríguez, María Teresa. 2003. *Ritual, identidad y procesos étnicos en la sierra de Zongolica, Veracruz.* Mexico City: Centro de Investigaciones y Estudios Superiores en Antropología Social.

Roque, Hilario Concepción. 2012. *Historia cuicateca.* Mérida: Fondo Institucional de Fomento Regional para el Desarrollo Científico, Tecnológico y de Innovación.

Rzedowski, Jerzy. 1978. *La vegetación de México.* Mexico City: Editorial Limusa Wiley.

Rzedowski, Jerzy, and Graciela Calderón de Rzedowski. 2013. "Datos para la apreciación de la flora fanerogámica del bosque tropical caducifolio de México." *Acta Botánica Mexicana* 102:1–23.

Rzedowski, Jerzy, R. Medina-Lémus, and Graciela Calderón de Rzedowski. 2004. "Las especies de *Bursera* (Burseraceae) en la cuenca superior del Río Papaloapan (México)." *Acta Botánica Mexicana* 66:23–151.

Scarborough, Vernon. 2000. "Hydrology." In *Archaeology of Ancient Mexico and Central America: An Encyclopedia,* edited by Susan Toby Evans and David L. Webster, 352–56. London: Routledge.

Schele, Linda, and David Freidel. 1990. *A Forest of Kings*. New York: HarperCollins.

Sedlock, Richard, Fernando Ortega-Gutiérrez, and Robert Speed. 1993. *Technostratigraphic Terranes and Tectonic Evolution of Mexico*. Special Paper 278. Boulder, Colo.: Geological Society of America.

Shimada, Izumi. 1999. "Evolution of Andean Diversity: Regional Formations (500 B.C.E–C.E. 600)." In *The Cambridge History of the Native Peoples of the Americas*, vol. 3, pt. 1, edited by Frank Saloman and Stuart B. Press, 350–517. Cambridge: Cambridge University Press.

Sisson, Edward. 1973. *First Annual Report of the Coxcatlán Project*. Andover, Mass.: Phillips Academy.

Sisson, Edward. 1997. "La muerte en el cacicazgo de Coxcatlán." In *Simpósium internacional Tehuacán y su entorno: Balance y perspectivas*, edited by Eréndira de la Lama, 129–49. Mexico City: INAH.

Smith, C. Earle. 1965a. "Agriculture, Tehuacán Valley." *Fieldiana: Botany* 31:49–100.

Smith, C. Earle. 1965b. "Flora, Tehuacán Valley." *Fieldiana: Botany* 31:101–44.

Smith, C. Earle. 1967. "Plant Remains." In *The Prehistory of the Tehuacán Valley*, vol. 1, *Environment and Subsistence*, edited by D. S. Byers, 220–55. Austin: University of Texas Press.

Smith, Charlotte A. 2002. "Concordant Change and Core-Periphery Dynamics: A Synthesis of Highland Mesoamerican Archaeological Survey Data." PhD diss., University of Georgia.

Smith, Michael E. 2005. "City Size in Late Post-Classic Mesoamerica." *Journal of Urban History* 31 (4): 403–34.

Sosa V., J. A. De-Nova, and M. Vázquez-Cruz. 2018. "Evolutionary History of the Flora of Mexico: Dry Forests: Cradles and Museums of Endemism." *Journal of Systematics and Evolution* 56:523–36. https://doi.org/10.1111/jse.12416.

Sousa, M. 1969. "Las colecciones botánicas de C. A. Purpus en México." *University of California Publications in Botany* 51:1–36.

Spencer, Charles S. 2003. "War and Early State Formation in Oaxaca, Mexico." *Proceeding of the National Academy of Sciences of the United States of America* 100:11185–87.

Spencer, Charles S., and Elsa M. Redmond. 1997. *Archaeology of the Cañada de Cuicatlán, Oaxaca*. American Museum of Natural History, Anthropological Papers, Number 80. New York: American Museum of Natural History.

Spencer, Charles, and Elsa Redmond. 2000. "Lightning and Jaguars: Iconography, Ideology, and Politics in Formative Cuicatlán, Oaxaca." In *Cultural Evolution: Contemporary Viewpoints*, edited by Gary M. Feinman and Linda Manzanilla, 145–76. New York: Kluwer Academic/Plenum.

Spencer, Charles S., and Elsa Redmond. 2001. "The Chronology of Conquest: Implications of New Radiocarbon Analyses from the Cañada de Cuicatlán, Oaxaca." *Latin American Antiquity* 12 (2): 182–201.

Spencer, Charles, and Elsa Redmond. 2004. "Primary State Formation in Mesoamerica." *Annual Review of Anthropology* 33:173–99.

Spores, Ronald. 1967. *The Mixtec Kings and Their People*. Norman: University of Oklahoma Press.

Spores, Ronald. 1969. "Settlement, Farming, Technology, and Environment in the Nochixtlán Valley." *Science* 166:557–69.

Spores, Ronald, and Andrew Balkansky. 2013. *The Mixtecs of Oaxaca: Ancient Times to the Present*. Norman: University of Oklahoma Press.

Starr, Greg. 2012. *Agaves: Living Sculptures for Landscapes and Containers*. Portland, Ore.: Timber Press.

Starr, Greg D., and Tristan J. Davis. 2019. "*Agave oteroi* (Asparagaceae/Agavoideae), a New Species from North-Central Oaxaca, Mexico." *Cactus and Succulent Journal* 91 (2): 134–43.

Suárez Cruz, Sergio. 1985. *Un entierro del Clásico superior en Cholula, Puebla*. Cuaderno de Trabajo, 6. Mexico City: Centro Regional Puebla, INAH, 1985.

Suárez, Jorge. 1983. *Las lenguas indígenas mesoamericanas*. Mexico City: Instituto Nacional Indigenista, Centro de Investigaciones y Estudios Superiores en Antropología Social.

Terraciano, Kevin. 1994. "Ñudzahui History: Mixtec Writing and Culture in Colonial Oaxaca." PhD diss., University of California, Los Angeles.

Terraciano, Kevin. 2002. *The Mixtecs of Colonial Oaxaca*. Palo Alto, Calif.: Stanford University Press.

Terzaghi, K. 1986. "There Were Giants on Earth in Those Days." In *Terzaghi Lectures, 1974–1982*, 219–20. Geotechnical Special Publication Series 1. New York: American Society of Civil Engineers.

Thomas, Hugh. 1993. *Conquest: Montezuma, Cortés, and the Fall of Old Mexico*. New York: Touchstone.

Trasviña, A., E. Barton, J. Brown, H. Vélez, and R. Smith. 1995. "Offshore Wind Forcing in the Gulf of Tehuantepec, Mexico: Asymmetric Circulation." *Journal of Geophysical Research* 100:20649–63.

Tucker, Tim, and Arturo Montero, eds. 2008. *Mapa de Cuauhtinchan II: Entre la ciencia y lo sagrado*. Mexico City: UNAM, Mesoamerican Research Foundation.

Valencia-Avalos, Susana, and Kevin C. Nixon. 2004. "Encinos." In *Biodiversidad de Oaxaca*, edited by Abisaí J. García-Mendoza, María de Jesús Ordóñez, and Miguel Briones-Salas, 219–24. Mexico City: Universidad Nacional Autónoma de México.

Valiente-Banuet, Alfonso, A. Bolongaro-Crevenna, O. Briones, E. Ezcurra, M. Rosas, H. Núñez, G. Barnard, and E. Vázquez. 1991. "Spatial Relationships between Cacti and Nurse Shrubs in a Semi-arid Environment in Central Mexico." *Journal of Vegetation Science* 2:15–20.

Valiente-Banuet, Alfonso, and Exequiel Ezcurra. 1991. "Shade as a Cause of the Association between the Cactus *Neobuxbaumia tetetzo* and the Nurse-Plant *Mimosa luisana* in the Tehuacán Valley, Mexico." *Journal of Ecology* 79:961–71.

Valiente-Banuet, Alfonso, Leonor Solís, Patricia Dávila, María del Coro Arizmendi, Carlos Silva Pereyra, José Ortega-Ramírez, Jacinto Treviño Carreón, Selene Rangel-Landa, and Alejandro Casas. 2009. *Guía de la vegetación del Valle de Tehuacán-Cuicatlán*. Mexico City: Impresora Continental de México.

Valiente-Banuet, Alfonso, F. Vite, and A. Zavala-Hurtado. 1991. "Interaction Between the Cactus *Neobuxbaumia tetetzo* and the Nurse Shrub *Mimosa luisana*." *Journal of Vegetation Science* 2:11–14.

Valverde, P. L., F. Vite, M. A. Pérez-Hernández, and J. A. Zavala-Hurtado. 2010. "Stem Tilting, Pseudocephalium Orientation, and Stem Allometry in *Cephalocereus columna-trajani* along a Short Latitudinal Gradient." *Plant Ecology* 188:17–27.

Van Doesburg, Bas. 2001. "The Codex Porfirio Díaz and the Map of Tutepetongo: The Curious Relationship between Pictography and Glosses in Oaxacan Screenfolds." *Ethnohistory* 48:403–32.

Villa R., Alfonso. 1934. *The Yaxuna-Cobá Causeway*. Contributions to American Archaeology 2. Washington, D.C.: Carnegie Institution of Washington.

Warriner, C., N. Robles García, R. Spores, and N. Tuross. 2012. "Disease, Demography, and Diet in Early Colonial New Spain: Investigation of a Sixteenth-Century Mixtec Cemetery at Teposculula Yucundaa." *Society for American Archaeology* 23:467–89.

Weeks, A., F. Zapata, S. K. Pell, D. C. Daly, J. D. Mitchell, and P. V. A. Fine. 2014. "To Move or to Evolve: Contrasting Patterns of Intercontinental Connectivity and Climatic Niche Evolution in 'Terebinthaceae' (Anacardiaceae and Burseraceae)." *Frontiers in Genetics* 5:409. https://doi.org/10.3389/fgene.2014.00409.

Weeks, A., D. C. Daly, and Beryl B. Simpson. 2005. "The Phylogenetic History and Biogeography of the Frankincense and Myrrh Family (Burseraceae) Based on Nuclear and Chloroplast Sequence Data." *Molecular Phylogenetics and Evolution* 35 (1): 85–101.

Whalen, M. E., and P. E. Minnis. 2009. *The Neighbors of Casas Grandes: Excavating Medio Period Communities of Northwest Chihuahua, Mexico*. Tucson: University of Arizona Press.

Williams, Eduardo. 2016. *La sal de la tierra*. 2nd ed. Morelia: El Colegio de Michoacán.

Williams, Robert. 2009. "Codex Zouche-Nuttall Pages 1–41: Narrative, Structure, Contents, and Chronologies." PhD diss., University of Texas, Austin.

Winter, Marcus. 2007. *Cerro de Las Minas: Arqueología de la Mixteca Baja*. Oaxaca: Centro INAH.

Wolf, Eric. 1959. *Sons of the Shaking Earth*. Chicago: University of Chicago Press.

Woodbury, Richard, and James Neely. 1972. "Water Control Systems of the Tehuacán Valley." In *The Prehistory of the Tehuacán Valley*, vol. 4, *Chronology and Irrigation*, edited by Frederick Johnson, 81–153. Austin: University of Texas Press.

INDEX

Please note: Page number in bold type indicate photograph

ABOUT THE AUTHORS

David Yetman is a research social scientist at the Southwest Center of the University of Arizona, as well as host and co-producer of the PBS travel/adventure series *In the Americas with David Yetman*. He is a celebrated storyteller who has called Arizona home for most of his life.

Alberto Búrquez is a researcher at the Instituto de Ecología, Universidad Nacional Autónoma de México. His widely cited research focuses on plant-animal interactions, biogeography, and the ecology of dry-land plants.

The Southwest Center Series

Jeffrey M. Banister, Editor

Ignaz Pfefferkorn, *Sonora: A Description of the Province*

Carl Lumholtz, *New Trails in Mexico*

Buford Pickens, *The Missions of Northern Sonora: A 1935 Field Documentation*

Gary Paul Nabhan, editor, *Counting Sheep: Twenty Ways of Seeing Desert Bighorn*

Eileen Oktavec, *Answered Prayers: Miracles and Milagros Along the Border*

Curtis M. Hinsley and David R. Wilcox, editors, *Frank Hamilton Cushing and the Hemenway Southwestern Archaeological Expedition, 1886–1889*, volume 1: *The Southwest in the American Imagination: The Writings of Sylvester Baxter, 1881–1899*

Lawrence J. Taylor and Maeve Hickey, *The Road to Mexico*

Donna J. Guy and Thomas E. Sheridan, editors, *Contested Ground: Comparative Frontiers on the Northern and Southern Edges of the Spanish Empire*

Julian D. Hayden, *The Sierra Pinacate*

Paul S. Martin, David Yetman, Mark Fishbein, Phil Jenkins, Thomas R. Van Devender, and Rebecca K. Wilson, editors, *Gentry's Rio Mayo Plants: The Tropical Deciduous Forest and Environs of Northwest Mexico*

W. J. McGee, *Trails to Tiburón: The 1894 and 1895 Field Diaries of W J McGee*, transcribed by Hazel McFeely Fontana, annotated and with an introduction by Bernard L. Fontana

Richard Stephen Felger, *Flora of the Gran Desierto and Río Colorado of Northwestern Mexico*

Donald Bahr, editor, *O'odham Creation and Related Events: As Told to Ruth Benedict in 1927 in Prose, Oratory, and Song by the Pimas William Blackwater, Thomas Vanyiko, Clara Ahiel, William Stevens, Oliver Wellington, and Kisto*

Dan L. Fischer, *Early Southwest Ornithologists, 1528–1900*

Thomas Bowen, editor, *Backcountry Pilot: Flying Adventures with Ike Russell*

Federico José María Ronstadt, *Borderman: Memoirs of Federico José María Ronstadt, edited by Edward F. Ronstadt*

Curtis M. Hinsley and David R. Wilcox, editors, *Frank Hamilton Cushing and the Hemenway Southwestern Archaeological Expedition, 1886–1889*, volume 2: *The Lost Itinerary of Frank Hamilton Cushing*

Neil Goodwin, *Like a Brother: Grenville Goodwin's Apache Years, 1928–1939*

Katherine G. Morrissey and Kirsten Jensen, editors, *Picturing Arizona: The Photographic Record of the 1930s*

Bill Broyles and Michael Berman, *Sunshot: Peril and Wonder in the Gran Desierto*

David W. Lazaroff, Philip C. Rosen, and Charles H. Lowe Jr., *Amphibians, Reptiles, and Their Habitats at Sabino Canyon*

David Yetman, *The Organ Pipe Cactus*

Gloria Fraser Giffords, *Sanctuaries of Earth, Stone, and Light: The Churches of Northern New Spain, 1530–1821*

David Yetman, *The Great Cacti: Ethnobotany and Biogeography*

John Messina, *Álamos, Sonora: Architecture and Urbanism in the Dry Tropics*

Laura L. Cummings, *Pachucas and Pachucos in Tucson: Situated Border Lives*

Bernard L. Fontana and Edward McCain, *A Gift of Angels: The Art of Mission San Xavier del Bac*

David A. Yetman, *The Ópatas: In Search of a Sonoran People*

Julian D. Hayden, *Field Man: The Life of a Desert Archaeologist*, edited by Bill Broyles and Diane Boyer

Bill Broyles, Gayle Harrison Hartmann, Thomas E. Sheridan, Gary Paul Nabhan, and Mary Charlotte Thurtle, *Last Water on the Devil's Highway: A Cultural and Natural History of Tinajas Altas*

Thomas E. Sheridan, *Arizona: A History, Revised Edition*

Richard S. Felger and Benjamin Theodore Wilder, *Plant Life of a Desert Archipelago: Flora of the Sonoran Islands in the Gulf of California*

David Burkhalter, *Baja California Missions: In the Footsteps of the Padres*

Guillermo Núñez Noriega, *Just Between Us: An Ethnography of Male Identity and Intimacy in Rural Communities of Northern Mexico*

Cathy Moser Marlett, *Shells on a Desert Shore: Mollusks in the Seri World*

Rebecca A. Carte, *Capturing the Landscapes of New Spain: Baltasar Obregón and the 1564 Ibarra Expedition*

Gary Paul Nabhan, editor, *Ethnobiology for the Future: Linking Cultural and Ecological Diversity*

James S. Griffith, *Saints, Statues, and Stories: A Folklorist Looks at the Religious Art of Sonora*

David Yetman, Alberto Búrquez, Kevin Hultine, and Michael Sanderson, with Frank S. Crosswhite, *The Saguaro Cactus: A Natural History*

Carolyn Niethammer, *A Desert Feast: Celebrating Tucson's Culinary Heritage*

Gary Paul Nabhan, editor, *The Nature of Desert Nature*

David Yetman, *Natural Landmarks of Arizona*

Michael Chiago Sr. and Amadeo M. Rea, *Michael Chiago: O'odham Lifeways Through Art*

David Wentworth Lazaroff, *Picturing Sabino: A Photographic History of a Southwestern Canyon*

David Yetman and Alberto Búrquez, *Mexico's Valleys of Cuicatlán and Tehuacán: From Deserts to Clouds*